Quick Reference for the

Electrical and Computer Engineering PE Exam

Second Edition

John A. Camara, PE

Professional Publications, Inc. • Belmont, CA

How to Locate Errata and Other Updates for This Book

At Professional Publications, we do our best to bring you error-free books. But when errors do occur, we want to make sure that you know about them so they cause as little confusion as possible.

A current list of known errata and other updates for this book is available on the PPI website at **www.ppi2pass.com**. From the website home page, click on "Errata." We update the errata page as often as necessary, so check in regularly. You will also find instructions for submitting suspected errata. We are grateful to every reader who takes the time to help us improve the quality of our books by pointing out an error.

QUICK REFERENCE FOR THE ELECTRICAL AND COMPUTER ENGINEERING PE EXAM
Second Edition

Current printing of this edition: 2

Printing History

edition number	printing number	update
1	8	Minor corrections.
2	1	New edition. Copyright update.
2	2	Minor corrections.

Printed in the United States of America

Professional Publications, Inc.
1250 Fifth Avenue, Belmont, CA 94002
(650) 593-9119
www.ppi2pass.com

Library of Congress Cataloging-in-Publication Data
Camara, John A., 1956-
    Quick reference for the electrical and computer engineering PE exam / John A. Camara. -- 2nd ed.
        p. cm.
    ISBN 1-888577-79-7
    1. Electric engineering--United States--Examinations--Study guides. 2. Electric engineering--Problems, exercises, etc. I. Title.

TA169.C36 2002
621.3'076--dc21
                                                    2002021993

Table of Contents

How to Use This Book

This book (and others in the Quick Reference series) was developed to help you minimize problem-solving time on the PE exam. *Quick Reference* is a consolidation of the most useful equations in the *Electrical Engineering Reference Manual*. Using *Quick Reference*, you won't need to wade through pages of descriptive text when all you actually need is a quick look at the formulas to remind you of your next solution step.

The idea is this: you study for the exam using (primarily) your *Reference Manual*, and you take the exam using (primarily) your *Quick Reference*.

Once you've studied and mastered the theory behind an exam topic, you're ready to tackle practice problems. Save time by using this *Quick Reference* right from the start, as you begin solving problems. This book follows the same order and uses the same nomenclature as the *Electrical Engineering Reference Manual*. Once you become familiar with the sequencing of subjects in the *Reference Manual*, you'll be at home with *Quick Reference*. As you progress in your understanding of topics, you'll find you can rely more and more on *Quick Reference* for rapid retrieval of equations—without needing to refer back to the *Reference Manual*.

There are also times during problem-solving when you do need access to two kinds of information simultaneously—formulas and data, or formulas and nomenclature, or formulas and theory. We've all experienced the frustration of having to work problems using a spare pencil, a calculator, and the water-bill envelope to serve as page markers in a single book. *Quick Reference* provides a convenient way to keep the equations you need in front of you, even as you may be flip-flopping back and forth between theory and data in your other references.

Quick Reference also provides a convenient place for neatnik engineers to add their own comments and reminders to equations, without having to mess up their primary references. We expect you to throw this book away after the exam, so go ahead and write in it.

Once you start incorporating *Quick Reference* into your problem-solving routine, we predict you won't want to return to the one-book approach. *Quick Reference* will save you precious time—and that's how to use this book.

Codes, Handbooks, and References

This edition of *Quick Reference for the Electrical Engineering PE Exam* is based on the following codes, standards, and references. The most current versions or editions available were used. However, the PE examination is not always based on the most currently available codes, as adoption by state and local controlling authorities often lags issuance by several years.

The PPI website (**www.ppi2pass.com**) provides the dates of the codes, standards, and regulations on which NCEES has announced the current exams are based. Use this information to decide which versions of these books should be part of your exam preparation.

The minimum recommended library for the electrical exam consists of this book, the *Electrical Engineering Reference Manual*, the current National Electric Code, a standard handbook of electrical engineering, and two textbooks that cover fundamental circuit theory (both electrical and electronic). Numerous textbooks covering basic electrical engineering topics are listed. These texts, or their equivalent (see the topics in brackets), should be used in preparation for the examination. Adequate preparation, not an extensive portable library, is the key to success.

Codes

Code of Federal Regulations, Title 47—Telecommunications, Ch. 73—"Radio Broadcast Rules." 47CFR73. U.S. Government.[1]

National Electric Code, NFPA 70. National Fire Protection Association.

National Electric Safety Code, NESC.

Standards Organizations of Interest:

ANSI: American National Standards Institute

EIA: Electronic Industries Alliance

FCC: Federal Communications Commission

IEEE: Institute of Electrical and Electronic Engineers

ISA: Instrument Society of America

ISO: International Organization for Standardization

NEMA: National Electrical Manufacturers Association

---

[1] The chapters within this title of the Code of Federal Regulations are known as the *FCC Rules*.

Handbooks

American Electrician's Handbook. Terrell Croft and Wilford I. Summers. McGraw-Hill.

CRC Materials Science and Engineering Handbook. James F. Shackelford, William Alexander, and Jun S. Park, eds. CRC Press, Inc.

Electronic Engineer's Handbook. Donald Christiansen, ed. McGraw-Hill.

McGraw-Hill Internetworking Handbook. Ed Taylor. McGraw-Hill.

National Electric Code Handbook. Mark W. Earley, Joseph V. Sheehan, and John M. Caloggero. National Fire Protection Association.

Standard Handbook for Electrical Engineers. Donald G. Fink and H. Wayne Beaty. McGraw-Hill.

The Communications Handbook. Jerry D. Gibson, ed. CRC Press, Inc.

The Computer Science and Engineering Handbook. Allen B. Tucker, Jr., ed. CRC Press, Inc.

References

CRC Standard Mathematical Tables. William H. Beyer, ed. CRC Press, Inc.

McGraw-Hill Dictionary of Scientific and Technical Terms. Sybil P. Parker, ed. McGraw-Hill.

Schaum's Outline Series (Electronics and Electrical Engineering). McGraw-Hill.

The Internet for Scientists and Engineers, Brian J. Thomas. SPIE Press & IEEE Press.

Texts

An Introduction to Digital and Analog Integrated Circuits and Applications. Sanjit K. Mitra. Harper & Row, Publishers. [Digital Circuit Fundamentals]

Applied Electromagnetics. Martin A. Plonus. McGraw-Hill. [Electromagnetic Theory]

Introduction to Computer Engineering. Taylor L. Booth. John Wiley & Sons. [Computer Design Basics]

Electrical Power Technology. Theodore Wildi. John Wiley & Sons. [Power Theory and Application]

Linear Circuits. M.E. Van Valkenburg and B.K. Kinariwala. Prentice-Hall, Inc. [AC/DC Fundamentals]

Microelectronics. Jacob Millman. McGraw-Hill. [Electronic Fundamentals]

For Instant Recall

Temperature Conversions

$$^\circ\text{F} = 32 + \tfrac{9}{5}\,^\circ\text{C}$$
$$^\circ\text{C} = \tfrac{5}{9}(^\circ\text{F} - 32)$$
$$^\circ\text{R} = \,^\circ\text{F} + 460$$
$$\text{K} = \,^\circ\text{C} + 273$$
$$\Delta^\circ\text{R} = \tfrac{9}{5}\Delta\text{K}$$
$$\Delta\text{K} = \tfrac{5}{9}\Delta^\circ\text{R}$$

Common SI Unit Conversion Factors

multiply	by	to obtain
AREA		
circular mil	506.7	square micrometer
ENERGY		
Btu (international)	1.0551	kilojoule
erg	0.1	microjoule
foot-pound	1.3558	joule
horsepower-hour	2.6485	megajoule
joule	0.7376	foot-pound
	0.10197	meter-kilogram force
kilogram-calorie (international)	4.1868	kilojoule
kilojoule	0.9478	Btu
	0.2388	kilogram-calorie
kilowatt-hour	3.6	megajoule
megajoule	0.3725	horsepower-hour
	0.2778	kilowatt-hour
	0.009478	therm
meter-kilogram force	9.8067	joule
microjoule	10.0	erg
therm	105.506	megajoule
FORCE		
dyne	10.0	micronewton
kilogram force	9.8067	newton
kip	4448.2	newton
micronewton	0.1	dyne
newton	0.10197	kilogram force
	0.0002248	kip
	3.597	ounce force
	0.2248	pound force
ounce force	0.2780	newton
pound force	4.4482	newton

(continued)

PROFESSIONAL PUBLICATIONS, INC.

Common SI Unit Conversion Factors (*continued*)

multiply	by	to obtain
LENGTH		
angstrom	0.1	nanometer
foot	0.3048	meter
inch	25.4	millimeter
kilometer	0.6214	mile
	0.540	mile (nautical)
meter	3.2808	foot
	1.0936	yard
micrometer	1.0	micron
micron	1.0	micrometer
mil	0.0254	millimeter
mile	1.6093	kilometer
millimeter	0.0394	inch
	39.370	mil
nanometer	10.0	angstrom
yard	0.9144	meter
MASS (weight)		
kilogram	2.2046	pound mass
	0.068522	slug
	0.0009842	ton (long—2240 lbm)
	0.001102	ton (short—2000 lbm)
pound mass	0.4536	kilogram
POWER		
Btu (international)/hr	0.2931	watt
foot-pound/sec	1.3558	watt
horsepower	0.7457	kilowatt
kilowatt	1.341	horsepower
	0.2843	ton of refrigeration
meter·kilogram force/sec	9.8067	watt
ton of refrigeration	3.517	kilowatt
watt	3.4122	Btu (international)/hr
	0.7376	foot-pound/sec
	0.10197	meter·kilogram force/sec
PRESSURE		
kilopascal	0.01	bar
	4.0219	inch, H_2O (20°C)
	0.2964	inch, Hg (20°C)
	0.0102	kilogram force/cm^2
	7.528	millimeter Hg (20°C)
	0.1450	pound force/in^2
	0.009869	standard atmosphere (760 torr)
	7.5006	torr
millimeter Hg (20°C)	0.13284	kilopascal
pound force/in^2	6.8948	kilopascal
standard atmosphere (760 torr)	101.325	kilopascal

(*continued*)

Common SI Unit Conversion Factors *(continued)*

multiply	by	to obtain
VOLUME (capacity)		
cubic centimeter	0.06102	cubic inch
cubic foot	28.3168	liter
cubic inch	16.3871	cubic centimeter
cubic meter	1.308	cubic yard
cubic yard	0.7646	cubic meter
gallon (U.S.)	3.785	liter
liter	0.2642	gallon (U.S.)
	2.113	pint (U.S. fluid)
	1.0567	quart (U.S. fluid)
	0.03531	cubic foot
milliliter	0.0338	ounce (U.S. fluid)
ounce (U.S. fluid)	29.574	milliliter
pint (U.S. fluid)	0.4732	liter
quart (U.S. fluid)	0.9464	liter

Common SI Unit Conversion Factors *(continued)*

PROFESSIONAL PUBLICATIONS, INC.

Mathematics

EERM Chapter 1
Systems of Units

Chapter, section, equation, figure, and table numbers correspond to EERM. For additional study material, go to the corresponding chapter and section number in EERM.

6. THE ENGLISH ENGINEERING SYSTEM

$$F \text{ in lbf} = \frac{(m \text{ in lbm}) \left(a \text{ in } \dfrac{\text{ft}}{\text{sec}^2} \right)}{g_c \text{ in} \dfrac{\text{lbm-ft}}{\text{lbf-sec}^2}} \qquad 1.6$$

g_c has a numerical value of 32.174.

7. FORMULAS AFFECTED BY INCONSISTENCY

- *kinetic energy*

$$E = \frac{mv^2}{2g_c} \quad [\text{in ft-lbf}] \qquad 1.7$$

- *potential energy*

$$E = \frac{mgz}{g_c} \quad [\text{in ft-lbf}] \qquad 1.8$$

EERM Chapter 2
Energy, Work, and Power

Chapter, section, equation, figure, and table numbers correspond to EERM. For additional study material, go to the corresponding chapter and section number in EERM.

3. WORK

$$W_{\text{constant force}} = \mathbf{F} \cdot \mathbf{s} = Fs \cos \phi \quad [\text{linear systems}] \qquad 2.6$$

$$W_{\text{constant torque}} = \mathbf{T} \cdot \boldsymbol{\theta} = Fr\theta \, \cos \phi$$

$$[\text{rotational systems}] \qquad 2.7$$

4. POTENTIAL ENERGY OF A MASS

$$E_{\text{potential}} = mgh \qquad 2.11(a)$$

5. KINETIC ENERGY OF A MASS

$$E_{\text{kinetic}} = \tfrac{1}{2}mv^2 \qquad 2.13(a)$$

$$E_{\text{rotational}} = \tfrac{1}{2}I\omega^2 \qquad 2.15(a)$$

6. SPRING ENERGY

$$E_{\text{spring}} = \tfrac{1}{2}k\delta^2 \qquad 2.16$$

7. PRESSURE ENERGY OF A MASS

$$E_{\text{flow}} = \frac{mp}{\rho} = mpv \qquad 2.17$$

8. INTERNAL ENERGY OF A MASS

$$U_2 - U_1 = Q \qquad 2.18$$

$$Q = mc\,\Delta T \qquad 2.19$$

11. POWER

$$P = \frac{W}{\Delta t} \qquad 2.25$$

$$P = Fv \quad [\text{linear systems}] \qquad 2.26$$

$$P = T\omega \quad [\text{rotational systems}] \qquad 2.27$$

EERM Chapter 4
Algebra

> Chapter, section, equation, figure, and table numbers correspond to EERM. For additional study material, go to the corresponding chapter and section number in EERM.

9. ROOTS OF QUADRATIC EQUATIONS

A *quadratic equation* is an equation of the general form $ax^2 + bx + c = 0$ [$a \neq 0$].

$$x_1, x_2 = \frac{-b \pm \sqrt{b^2 - 4ac}}{2a} \qquad 4.17$$

- If $(b^2 - 4ac) > 0$, the roots are real and unequal.
- If $(b^2 - 4ac) = 0$, the roots are real and equal. This is known as a *double root*.
- If $(b^2 - 4ac) < 0$, the roots are complex and unequal.

13. RULES FOR EXPONENTS AND RADICALS

$$b^0 = 1 \qquad [b \neq 0] \qquad 4.20$$

$$\left(\frac{a}{b}\right)^n = \frac{a^n}{b^n} \qquad [b \neq 0] \qquad 4.23$$

$$(ab)^n = a^n b^n \qquad 4.24$$

$$b^{\frac{m}{n}} = \sqrt[n]{b^m} = \left(\sqrt[n]{b}\right)^m \qquad 4.25$$

$$\sqrt[n]{b} = b^{\frac{1}{n}} \qquad 4.29$$

15. LOGARITHM IDENTITIES

$$\log_b(b) = 1 \qquad 4.34$$

$$\log_b(1) = 0 \qquad 4.35$$

$$\log(x^a) = a \log(x) \qquad 4.37$$

$$\log(xy) = \log(x) + \log(y) \qquad 4.41$$

$$\log\left(\frac{x}{y}\right) = \log(x) - \log(y) \qquad 4.42$$

$$\log_a(x) = \log_b(x) \log_a(b) \qquad 4.43$$

16. PARTIAL FRACTIONS

$H(x)$ is a proper polynomial fraction of the form $P(x)/Q(x)$.

$$H(x) = \frac{P(x)}{Q(x)} = \frac{u_1}{v_1} + \frac{u_2}{v_2} + \frac{u_3}{v_3} + \cdots \qquad 4.46$$

case 1: $Q(x)$ factors into n different linear terms.

$$Q(x) = (x - a_1)(x - a_2) \cdots (x - a_n) \qquad 4.47$$

Then,

$$H(x) = \sum_{i=1}^{n} \frac{A_i}{x - a_i} \qquad 4.48$$

case 2: $Q(x)$ factors into n identical linear terms.

$$Q(x) = (x - a)(x - a) \cdots (x - a) \qquad 4.49$$

Then,

$$H(x) = \sum_{i=1}^{n} \frac{A_i}{(x - a)^i} \qquad 4.50$$

case 3: $Q(x)$ factors into n different quadratic terms, $x^2 + p_i x + q_i$. Then,

$$H(x) = \sum_{i=1}^{n} \frac{A_i x + B_i}{x^2 + p_i x + q_i} \qquad 4.51$$

case 4: $Q(x)$ factors into n identical quadratic terms, $x^2 + px + q$. Then,

$$H(x) = \sum_{i=1}^{n} \frac{A_i x + B_i}{(x^2 + px + q)^i} \qquad 4.52$$

Once the general forms of the partial fractions have been determined from inspection, the *method of undetermined coefficients* is used.

18. COMPLEX NUMBERS

- *rectangular* or *trigonometric form*

$$\mathbf{Z} = a + bi$$

- *exponential form*

$$a + bi = re^{i\theta} \qquad 4.53$$

$$r = \text{mod}(\mathbf{Z}) = \sqrt{\mathbf{a^2 + b^2}} \qquad 4.54$$

$$\theta = \text{arg}(\mathbf{Z}) = \arctan\left(\frac{b}{a}\right) \qquad 4.55$$

- *phasor form (polar form)*

$$\mathbf{Z} = r \angle \theta \qquad 4.56$$

- *rectangular form*

$$a = r\cos\theta \qquad \text{4.57}$$

$$b = r\sin\theta \qquad \text{4.58}$$

$$\mathbf{Z} = a + bi = r\cos\theta + ir\sin\theta$$

$$= r(\cos\theta + i\sin\theta) \qquad \text{4.59}$$

- *cis form*

$$a + bi = r(\cos\theta + i\sin\theta) = r\operatorname{cis}\theta \qquad \text{4.60}$$

- *Euler's equation*

$$e^{i\theta} = \cos\theta + i\sin\theta \qquad \text{4.61}$$

- *related expressions*

$$e^{-i\theta} = \cos\theta - i\sin\theta \qquad \text{4.62}$$

$$\cos\theta = \frac{e^{i\theta} + e^{-i\theta}}{2} \qquad \text{4.63}$$

$$\sin\theta = \frac{e^{i\theta} - e^{-i\theta}}{2i} \qquad \text{4.64}$$

EERM Chapter 5
Linear Algebra

> Chapter, section, equation, figure, and table numbers correspond to EERM. For additional study material, go to the corresponding chapter and section number in EERM.

5. DETERMINANTS

$$\mathbf{A} = \begin{bmatrix} a & b \\ c & d \end{bmatrix}$$

$$|\mathbf{A}| = \begin{vmatrix} a & b \\ c & d \end{vmatrix} = ad - bc \qquad \text{5.3}$$

$$\mathbf{A} = \begin{bmatrix} a & b & c \\ d & e & f \\ g & h & i \end{bmatrix}$$

augmented $\mathbf{A} = \begin{bmatrix} a & b & c \\ d & e & f \\ g & h & i \end{bmatrix}$ (with $+\,+\,+\,-\,-\,-$ diagonal products) $\qquad$ 5.4

$$|\mathbf{A}| = aei + bfg + cdh - gec - hfa - idb \qquad \text{5.5}$$

$$\mathbf{A} = \begin{bmatrix} a & b & c \\ d & e & f \\ g & h & i \end{bmatrix}$$

$$|\mathbf{A}| = a\begin{vmatrix} e & f \\ h & i \end{vmatrix} - d\begin{vmatrix} b & c \\ h & i \end{vmatrix} + g\begin{vmatrix} b & c \\ e & f \end{vmatrix} \qquad \text{5.6}$$

6. MATRIX ALGEBRA

- *commutative law of addition*

$$\mathbf{A} + \mathbf{B} = \mathbf{B} + \mathbf{A} \qquad \text{5.7}$$

- *associative law of addition*

$$\mathbf{A} + (\mathbf{B} + \mathbf{C}) = (\mathbf{A} + \mathbf{B}) + \mathbf{C} \qquad \text{5.8}$$

- *associative law of multiplication*

$$(\mathbf{AB})\mathbf{C} = \mathbf{A}(\mathbf{BC}) \qquad \text{5.9}$$

- *left distributive law*

$$\mathbf{A}(\mathbf{B} + \mathbf{C}) = \mathbf{AB} + \mathbf{AC} \qquad \text{5.10}$$

- *right distributive law*

$$(\mathbf{B} + \mathbf{C})\mathbf{A} = \mathbf{BA} + \mathbf{CA} \qquad \text{5.11}$$

- *scalar multiplication*

$$k(\mathbf{AB}) = (k\mathbf{A})\mathbf{B} = \mathbf{A}(k\mathbf{B}) \qquad \text{5.12}$$

13. WRITING SIMULTANEOUS LINEAR EQUATIONS IN MATRIX FORM

$$a_{11}x_1 + a_{12}x_2 = b_1$$
$$a_{21}x_1 + a_{22}x_2 = b_2$$

$$\begin{bmatrix} a_{11} & a_{12} \\ a_{21} & a_{22} \end{bmatrix} \begin{bmatrix} x_1 \\ x_2 \end{bmatrix} = \begin{bmatrix} b_1 \\ b_2 \end{bmatrix}$$

$$\mathbf{AX} = \mathbf{B}$$

$\mathbf{A}$ is known as the *coefficient matrix*, $\mathbf{X}$ as the *variable matrix*, and $\mathbf{B}$ as the *constant matrix*.

Table 5.1 *Solution Existence Rules for Simultaneous Equations*

	$\mathbf{B} = 0$	$\mathbf{B} \neq 0$
$\lvert\mathbf{A}\rvert = 0$	infinite number of solutions (linearly dependent equations)	either an infinite number of solutions or no solution at all
$\lvert\mathbf{A}\rvert \neq 0$	trivial solution only $(x_i = 0)$	unique nonzero solution

14. SOLVING SIMULTANEOUS LINEAR EQUATIONS

$$x_1 = \frac{|\mathbf{A}_1|}{|\mathbf{A}|} \qquad 5.23$$

$$x_2 = \frac{|\mathbf{A}_2|}{|\mathbf{A}|} \qquad 5.24$$

EERM Chapter 6
Vectors

Chapter, section, equation, figure, and table numbers correspond to EERM. For additional study material, go to the corresponding chapter and section number in EERM.

2. VECTORS IN *n*-SPACE

$$|\mathbf{V}| = \sqrt{(x_2 - x_1)^2 + (y_2 - y_1)^2} \qquad 6.1$$

$$\phi = \arctan\left(\frac{y_2 - y_1}{x_2 - x_1}\right) \qquad 6.2$$

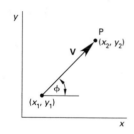

Figure 6.1 *Vector in Two-Dimensional Space*

$$V_x = |\mathbf{V}| \cos \phi_x \qquad 6.3$$

$$V_y = |\mathbf{V}| \cos \phi_y \qquad 6.4$$

$$V_z = |\mathbf{V}| \cos \phi_z \qquad 6.5$$

$$|\mathbf{V}| = \sqrt{V_x^2 + V_y^2 + V_z^2} \qquad 6.6$$

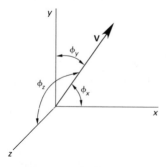

Figure 6.2 *Direction Angles of a Vector*

- *direction cosines*

$$\cos^2 \phi_x + \cos^2 \phi_y + \cos^2 \phi_z = 1 \qquad 6.7$$

3. UNIT VECTORS

$$\mathbf{V} = |\mathbf{V}|\mathbf{a} = V_x\mathbf{i} + V_y\mathbf{j} + V_z\mathbf{k} \qquad 6.8$$

$$\mathbf{a} = \frac{\mathbf{V}}{|\mathbf{V}|} = \frac{V_x\mathbf{i} + V_y\mathbf{j} + V_z\mathbf{k}}{\sqrt{V_x^2 + V_y^2 + V_z^2}} \qquad 6.9$$

4. VECTOR REPRESENTATION

- *rectangular form*

$$\mathbf{A} \equiv A_x\mathbf{i} + A_y\mathbf{j} + A_z\mathbf{k} \quad \text{[three dimensions]}$$

- *phasor form (polar form)*

$$\mathbf{A} \equiv |\mathbf{A}|\angle\phi = A\angle\phi$$

6. VECTOR ADDITION

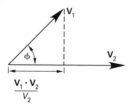

Figure 6.4 *Addition of Two Vectors*

7. MULTIPLICATION BY A SCALAR

$$c\mathbf{V} = c|\mathbf{V}|\,\mathbf{a} = cV_x\mathbf{i} + cV_y\mathbf{j} + cV_z\mathbf{k} \qquad 6.17$$

$$c(\mathbf{V}_1 + \mathbf{V}_2) = c\mathbf{V}_1 + c\mathbf{V}_2 \qquad 6.18$$

8. VECTOR DOT PRODUCT

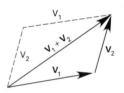

Figure 6.5 *Vector Dot Product*

$$\mathbf{V}_1 \cdot \mathbf{V}_2 = |\mathbf{V}_1||\mathbf{V}_2| \cos \phi$$

$$= V_{1x}V_{2x} + V_{1y}V_{2y} + V_{1z}V_{2z} \qquad 6.21$$

$$\cos \phi = \frac{V_{1x}V_{2x} + V_{1y}V_{2y} + V_{1z}V_{2z}}{|\mathbf{V}_1||\mathbf{V}_2|} \qquad 6.22$$

The dot product can be used to determine whether a vector is a unit vector and to show that two vectors are orthogonal (perpendicular). For two non-null orthogonal vectors,

$$\mathbf{V}_1 \cdot \mathbf{V}_2 = 0 \qquad 6.23$$

For any unit vector, **u**,

$$\mathbf{u} \cdot \mathbf{u} = 1 \qquad 6.24$$

9. VECTOR CROSS PRODUCT

$$\mathbf{V}_1 \times \mathbf{V}_2 = \begin{vmatrix} \mathbf{i} & V_{1x} & V_{2x} \\ \mathbf{j} & V_{1y} & V_{2y} \\ \mathbf{k} & V_{1z} & V_{2z} \end{vmatrix} \qquad 6.31$$

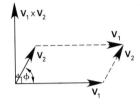

Figure 6.6 *Vector Cross Product*

10. MIXED TRIPLE PRODUCT

$$\mathbf{V}_1 \cdot (\mathbf{V}_2 \times \mathbf{V}_3) = \begin{vmatrix} V_{1x} & V_{1y} & V_{1z} \\ V_{2x} & V_{2y} & V_{2z} \\ V_{3x} & V_{3y} & V_{3z} \end{vmatrix} \qquad 6.40$$

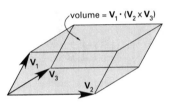

Figure 6.7 *Vector Mixed Triple Product*

12. VECTOR FUNCTIONS

$$\frac{d\mathbf{V}(t)}{dt} = \left(\frac{dV_x}{dt}\right)\mathbf{i} + \left(\frac{dV_y}{dt}\right)\mathbf{j} + \left(\frac{dV_z}{dt}\right)\mathbf{k} \qquad 6.44$$

$$\int \mathbf{V}(t)\,dt = \mathbf{i}\int V_x\,dt + \mathbf{j}\int V_y\,dt + \mathbf{k}\int V_z\,dt \qquad 6.45$$

EERM Chapter 7
Trigonometry

Chapter, section, equation, figure, and table numbers correspond to EERM. For additional study material, go to the corresponding chapter and section number in EERM.

1. DEGREES AND RADIANS

multiply	by	to obtain
radians	$\dfrac{180}{\pi}$	degrees
degrees	$\dfrac{\pi}{180}$	radians

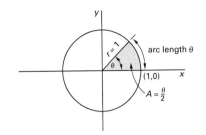

Figure 7.1 *Radians and Area of Unit Circle*

4. RIGHT TRIANGLES

- *Pythagorean theorem*

$$x^2 + y^2 = r^2 \qquad 7.1$$

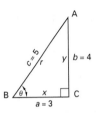

Figure 7.5 *3:4:5 Right Triangle*

5. CIRCULAR TRANSCENDENTAL FUNCTIONS

$$\text{sine: } \sin\theta = \frac{y}{r} = \frac{\text{opposite}}{\text{hypotenuse}} \qquad 7.2$$

$$\text{cosine: } \cos\theta = \frac{x}{r} = \frac{\text{adjacent}}{\text{hypotenuse}} \qquad 7.3$$

$$\text{tangent: } \tan\theta = \frac{y}{x} = \frac{\text{opposite}}{\text{adjacent}} \qquad 7.4$$

$$\text{cotangent: } \cot\theta = \frac{x}{y} = \frac{\text{adjacent}}{\text{opposite}} \qquad 7.5$$

$$\text{secant: } \sec\theta = \frac{r}{x} = \frac{\text{hypotenuse}}{\text{adjacent}} \qquad 7.6$$

$$\text{cosecant: } \csc\theta = \frac{r}{y} = \frac{\text{hypotenuse}}{\text{opposite}} \qquad 7.7$$

$$\cot\theta = \frac{1}{\tan\theta} \qquad 7.8$$

$$\sec\theta = \frac{1}{\cos\theta} \qquad 7.9$$

$$\csc\theta = \frac{1}{\sin\theta} \qquad 7.10$$

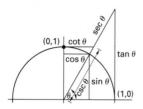

Figure 7.6 *Trigonometric Functions in a Unit Circle*

6. SMALL ANGLE APPROXIMATIONS

$$\sin\theta \approx \tan\theta \approx \theta\Big|_{\theta<10° \ (0.175 \ \text{rad})} \qquad 7.11$$

$$\cos\theta \approx 1\Big|_{\theta<5° \ (0.0873 \ \text{rad})} \qquad 7.12$$

10. TRIGONOMETRIC IDENTITIES

$$\sin^2\theta + \cos^2\theta = 1 \qquad 7.14$$

- *two-angle formulas*

$$\sin(\theta \pm \phi) = \sin\theta\cos\phi \pm \cos\theta\sin\phi \qquad 7.21$$

$$\cos(\theta \pm \phi) = \cos\theta\cos\phi \mp \sin\theta\sin\phi \qquad 7.22$$

14. GENERAL TRIANGLES

- *law of sines*

$$\frac{\sin A}{a} = \frac{\sin B}{b} = \frac{\sin C}{c} \qquad 7.57$$

- *law of cosines*

$$a^2 = b^2 + c^2 - 2bc\cos A \qquad 7.58$$

16. SOLID ANGLES

- *solid angle, ω*

$$\omega = \frac{\text{surface area}}{r^2} \qquad 7.67$$

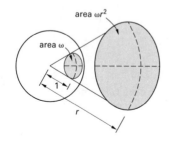

Figure 7.11 *Solid Angle*

EERM Chapter 8
Analytic Geometry

> Chapter, section, equation, figure, and table numbers correspond to EERM. For additional study material, go to the corresponding chapter and section number in EERM.

7. COORDINATE SYSTEMS

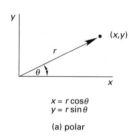

$x = r\cos\theta$
$y = r\sin\theta$

(a) polar

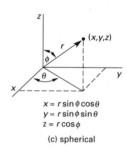

$x = r\cos\theta$
$y = r\sin\theta$
$z = z$

(b) cylindrical

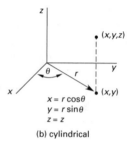

$x = r\sin\phi\cos\theta$
$y = r\sin\phi\sin\theta$
$z = r\cos\phi$

(c) spherical

Figure 8.5 *Different Coordinate Systems*

10. STRAIGHT LINES

- *general form*

$$Ax + By + C = 0 \qquad 8.8$$
$$A = -mB \qquad 8.9$$
$$B = \frac{-C}{b} \qquad 8.10$$
$$C = -aA = -bB \qquad 8.11$$

- *slope-intercept form*

$$y = mx + b \qquad 8.12$$
$$m = \frac{-A}{B} = \tan\theta = \frac{y_2 - y_1}{x_2 - x_1} \qquad 8.13$$

$$b = \frac{-C}{B}$$ 8.14

$$a = \frac{-C}{A}$$ 8.15

- *point-slope form*

$$y - y_1 = m(x - x_1)$$ 8.16

- *intercept form*

$$\frac{x}{a} + \frac{y}{b} = 1$$ 8.17

- *two-point form*

$$\frac{y - y_1}{x - x_1} = \frac{y_2 - y_1}{x_2 - x_1}$$ 8.18

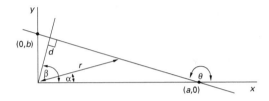

Figure 8.8 *Straight Line*

11. DIRECTION NUMBERS, ANGLES, AND COSINES

Given a directed line from (x_1, y_1, z_1) to (x_2, y_2, z_2), the direction numbers are

$$L = x_2 - x_1$$ 8.21

$$M = y_2 - y_1$$ 8.22

$$N = z_2 - z_1$$ 8.23

The distance between two points is

$$d = \sqrt{L^2 + M^2 + N^2}$$ 8.24

The direction cosines are

$$\cos \alpha = \frac{L}{d}$$ 8.25

$$\cos \beta = \frac{M}{d}$$ 8.26

$$\cos \gamma = \frac{N}{d}$$ 8.27

The direction angles are

$$\alpha = \arccos\left(\frac{L}{d}\right)$$ 8.29

$$\beta = \arccos\left(\frac{M}{d}\right)$$ 8.30

$$\gamma = \arccos\left(\frac{N}{d}\right)$$ 8.31

Line **R** would be defined as

$$\mathbf{R} = \mathbf{i}\cos \alpha + \mathbf{j}\cos \beta + \mathbf{k}\cos \gamma$$ 8.32

$$\mathbf{R} = L\mathbf{i} + M\mathbf{j} + N\mathbf{k}$$ 8.33

14. DISTANCES BETWEEN GEOMETRIC FIGURES

Between two points in (x, y, z) format,

$$d = \sqrt{(x_2 - x_1)^2 + (y_2 - y_1)^2 + (z_2 - z_1)^2}$$ 8.48

17. CIRCLE

- *general form*

$$Ax^2 + Ay^2 + Dx + Ey + F = 0$$ 8.66

- *center-radius form*

$$(x - h)^2 + (y - k)^2 = r^2$$ 8.67

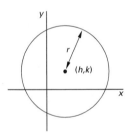

Figure 8.12 *Circle*

21. SPHERE

- *general equation*

$$Ax^2 + Ay^2 + Az^2 + Bx + Cy + Dz + E = 0$$ 8.104

- *sphere centered at (h, k, l) with radius r*

$$(x - h)^2 + (y - k)^2 + (z - l)^2 = r^2$$ 8.105

EERM Chapter 9
Differential Calculus

> Chapter, section, equation, figure, and table numbers correspond to EERM. For additional study material, go to the corresponding chapter and section number in EERM.

2. ELEMENTARY DERIVATIVE OPERATIONS

$$\mathbf{D}k = 0$$ 9.2

$$\mathbf{D}x^n = nx^{n-1}$$ 9.3

$$\mathbf{D}\ln x = \frac{1}{x} \qquad\qquad 9.4$$

$$\mathbf{D}e^{ax} = ae^{ax} \qquad\qquad 9.5$$

$$\mathbf{D}\sin x = \cos x \qquad\qquad 9.6$$

$$\mathbf{D}\cos x = -\sin x \qquad\qquad 9.7$$

$$\mathbf{D}\sinh x = \cosh x \qquad\qquad 9.18$$

$$\mathbf{D}\cosh x = \sinh x \qquad\qquad 9.19$$

$$\mathbf{D}kf(x) = k\mathbf{D}f(x) \qquad\qquad 9.24$$

$$\mathbf{D}(f(x) \pm g(x)) = \mathbf{D}f(x) \pm \mathbf{D}g(x) \qquad 9.25$$

$$\mathbf{D}(f(x) \cdot g(x)) = f(x)\mathbf{D}g(x) + g(x)\mathbf{D}f(x) \qquad 9.26$$

$$\mathbf{D}\left(\frac{f(x)}{g(x)}\right) = \frac{g(x)\mathbf{D}f(x) - f(x)\mathbf{D}g(x)}{(g(x))^2} \qquad 9.27$$

$$\mathbf{D}(f(x))^n = n(f(x))^{n-1}\mathbf{D}f(x) \qquad 9.28$$

$$\mathbf{D}f(g(x)) = \mathbf{D}_g f(g)\mathbf{D}_x g(x) \qquad 9.29$$

3. CRITICAL POINTS

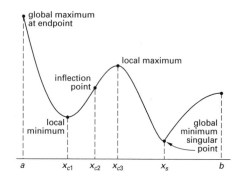

Figure 9.2 *Extreme and Inflection Points*

$$f'(x_c) = 0 \text{ at any extreme point, } x_c \qquad 9.30$$

$$f''(x_c) < 0 \text{ at a maximum point} \qquad 9.31$$

$$f''(x_c) > 0 \text{ at a minimum point} \qquad 9.32$$

$$f''(x_c) = 0 \text{ at an inflection point} \qquad 9.33$$

8. GRADIENT VECTOR

$$\nabla f(x,y,z) = \left(\frac{\partial f(x,y,z)}{\partial x}\right)\mathbf{i} + \left(\frac{\partial f(x,y,z)}{\partial y}\right)\mathbf{j}$$
$$+ \left(\frac{\partial f(x,y,z)}{\partial z}\right)\mathbf{k} \qquad 9.37$$

10. NORMAL LINE VECTOR

$$\mathbf{N} = \left.\frac{\partial f(x,y,z)}{\partial x}\right|_{P_0}\mathbf{i} + \left.\frac{\partial f(x,y,z)}{\partial y}\right|_{P_0}\mathbf{j}$$
$$+ \left.\frac{\partial f(x,y,z)}{\partial z}\right|_{P_0}\mathbf{k} \qquad 9.43$$

11. DIVERGENCE OF A VECTOR FIELD

$$\mathbf{F} = P(x,y,z)\mathbf{i} + Q(x,y,z)\mathbf{j} + R(x,y,z)\mathbf{k} \quad 9.44$$

$$\operatorname{div}\mathbf{F} = \frac{\partial P}{\partial x} + \frac{\partial Q}{\partial y} + \frac{\partial R}{\partial z} \qquad 9.45$$

$$\operatorname{div}\mathbf{F} = \nabla \cdot \mathbf{F} \qquad 9.46$$

$$\nabla = \frac{\partial}{\partial x}\mathbf{i} + \frac{\partial}{\partial y}\mathbf{j} + \frac{\partial}{\partial z}\mathbf{k} \qquad 9.47$$

12. CURL OF A VECTOR FIELD

$$\mathbf{F} = P(x,y,z)\mathbf{i} + Q(x,y,z)\mathbf{j} + R(x,y,z)\mathbf{k} \quad 9.48$$

$$\operatorname{curl}\mathbf{F} = \left(\frac{\partial R}{\partial y} - \frac{\partial Q}{\partial z}\right)\mathbf{i} + \left(\frac{\partial P}{\partial z} - \frac{\partial R}{\partial x}\right)\mathbf{j}$$
$$+ \left(\frac{\partial Q}{\partial x} - \frac{\partial P}{\partial y}\right)\mathbf{k} \qquad 9.51$$

$$\operatorname{curl}\mathbf{F} = \nabla \times \mathbf{F}$$
$$= \begin{vmatrix} \mathbf{i} & \mathbf{j} & \mathbf{k} \\ \dfrac{\partial}{\partial x} & \dfrac{\partial}{\partial y} & \dfrac{\partial}{\partial z} \\ P(x,y,z) & Q(x,y,z) & R(x,y,z) \end{vmatrix} \quad 9.52$$

14. COMMON SERIES APPROXIMATIONS

$$\sin x \approx x - \frac{x^3}{3!} + \frac{x^5}{5!} - \frac{x^7}{7!} + \cdots$$
$$+ (-1)^n \frac{x^{2n+1}}{(2n+1)!} \qquad 9.56$$

$$\cos x \approx 1 - \frac{x^2}{2!} + \frac{x^4}{4!} - \frac{x^6}{6!} + \cdots + (-1)^n \frac{x^{2n}}{(2n)!} \quad 9.57$$

$$e^x \approx 1 + x + \frac{x^2}{2!} + \frac{x^3}{3!} + \cdots + \frac{x^n}{n!} \qquad 9.60$$

$$\ln(1+x) \approx x - \frac{x^2}{2} + \frac{x^3}{3} - \frac{x^4}{4} + \cdots + (-1)^{n+1}\frac{x^n}{n} \quad 9.61$$

$$\frac{1}{1-x} \approx 1 + x + x^2 + x^3 + \cdots + x^n \qquad 9.62$$

EERM Chapter 10
Integral Calculus

Chapter, section, equation, figure, and table numbers correspond to EERM. For additional study material, go to the corresponding chapter and section number in EERM.

2. ELEMENTARY OPERATIONS

C and k represent constants. $f(x)$ and $g(x)$ are functions of x.

$$\int k\,dx = kx + C \qquad\qquad 10.2$$

$$\int x^m\,dx = \frac{x^{m+1}}{m+1} + C \qquad [m \neq -1] \quad 10.3$$

$$\int \frac{1}{x}dx = \ln|x| + C \qquad\qquad 10.4$$

$$\int e^{kx}dx = \frac{e^{kx}}{k} + C \qquad\qquad 10.5$$

$$\int xe^{kx}dx = \frac{e^{kx}(kx-1)}{k^2} + C \qquad 10.6$$

$$\int \ln x\,dx = x \ln x - x + C \qquad 10.8$$

$$\int \sin x\,dx = -\cos x + C \qquad 10.9$$

$$\int \cos x\,dx = \sin x + C \qquad 10.10$$

$$\int \frac{dx}{k^2 + x^2} = \frac{1}{k}\arctan\frac{x}{k} + C \qquad 10.15$$

$$\int \sin^2 x\,dx = \frac{1}{2}x - \frac{1}{4}\sin 2x + C \qquad 10.18$$

$$\int \cos^2 x\,dx = \frac{1}{2}x + \frac{1}{4}\sin 2x + C \qquad 10.19$$

$$\int kf(x)dx = k\int f(x)dx \qquad 10.27$$

$$\int (f(x) + g(x))dx = \int f(x)dx + \int g(x)dx \qquad 10.28$$

$$\int \frac{f'(x)}{f(x)}dx = \ln|f(x)| + C \qquad 10.29$$

3. INTEGRATION BY PARTS

$$\int f(x)dg(x) = f(x)g(x) - \int g(x)df(x) + C \quad 10.31$$

8. AVERAGE VALUE

$$\text{average value} = \frac{1}{b-a}\int_a^b f(x)dx \qquad 10.34$$

9. AREA

$$A = \int_a^b (f_1(x) - f_2(x))dx \qquad 10.35$$

15. FOURIER SERIES

- *Fourier's theorem*

$$f(t) = \tfrac{1}{2}a_0 + a_1\cos\omega t + a_2\cos 2\omega t + \cdots$$
$$+ b_1\sin\omega t + b_2\sin 2\omega t + \cdots \qquad 10.47$$

a_0 can often be determined by inspection since it is the average value of the waveform.

- *natural (fundamental) frequency, ω*

$$\omega = \frac{2\pi}{T} \qquad\qquad 10.48$$

The time domain can be normalized to the radian scale.

$$f(t) = \tfrac{1}{2}a_0 + a_1\cos t + a_2\cos 2t + \cdots$$
$$+ b_1\sin t + b_2\sin 2t + \cdots \qquad 10.49$$

$$a_0 = \frac{1}{\pi}\int_0^{2\pi} f(t)dt$$
$$= \frac{2}{T}\int_0^T f(t)dt \qquad\qquad 10.50$$

$$a_n = \frac{1}{\pi}\int_0^{2\pi} f(t)\cos nt\,dt$$
$$= \frac{2}{T}\int_0^T f(t)\cos nt\,dt \qquad [n \geq 1] \quad 10.51$$

$$b_n = \frac{1}{\pi}\int_0^{2\pi} f(t)\sin nt\,dt$$
$$= \frac{2}{T}\int_0^T f(t)\sin nt\,dt \qquad [n \geq 1] \quad 10.52$$

EERM Chapter 11
Differential Equations

Chapter, section, equation, figure, and table numbers correspond to EERM. For additional study material, go to the corresponding chapter and section number in EERM.

2. HOMOGENEOUS, FIRST-ORDER LINEAR DIFFERENTIAL EQUATIONS WITH CONSTANT COEFFICIENTS

- *general form*

$$y' + ky = 0 \qquad\qquad 11.1$$

- *solution*

$$y = Ae^{rx} = Ae^{-kx} \qquad 11.2$$

3. FIRST-ORDER LINEAR DIFFERENTIAL EQUATIONS

$$y' + p(x)y = g(x) \qquad 11.4$$

- *integrating factor*

$$u(x) = \exp\left[\int p(x)dx\right] \qquad 11.5$$

- *closed-form solution*

$$y = \frac{1}{u(x)}\left[\int u(x)g(x)dx + C\right] \qquad 11.6$$

6. HOMOGENEOUS, SECOND-ORDER LINEAR DIFFERENTIAL EQUATIONS WITH CONSTANT COEFFICIENTS

$$y'' + k_1 y' + k_2 y = 0 \qquad 11.13$$
$$r^2 + k_1 r + k_2 = 0 \qquad 11.14$$

There are three cases. If the two roots of Eq. 11.14 are real and different, the solution is

$$y = A_1 e^{r_1 x} + A_2 e^{r_2 x} \qquad 11.15$$

If the two roots are real and the same, the solution is

$$y = A_1 e^{rx} + A_2 x e^{rx} \qquad 11.16$$
$$r = \frac{-k_1}{2} \qquad 11.17$$

If the two roots are imaginary, the solution is

$$y = A_1 e^{\alpha x} \cos \omega x + A_2 e^{\alpha x} \sin \omega x \qquad 11.18$$

In all three cases, A_1 and A_2 must be found from the two initial conditions.

7. NONHOMOGENEOUS DIFFERENTIAL EQUATIONS

- *forcing function, $f(x)$*

$$y'' + p(x)y' + q(x)y = f(x) \qquad 11.19$$

The *complementary solution*, y_c, solves the complementary (i.e., homogeneous) problem. The *particular solution*, y_p, is any specific solution to the nonhomogeneous Eq. 11.19 that is known or can be found.

$$y = y_c + y_p \qquad 11.20$$

Table 11.1 *Particular Solutions**

form of $f(x)$	form of y_p
$P_n(x) = a_0 x^n + a_1 x^{n-1} + \cdots + a_n$	$x^s(A_0 x^n + A_1 x^{n-1} + \cdots + A_n)$
$P_n(x)e^{\alpha x}$	$x^s(A_0 x^n + A_1 x^{n-1} + \cdots + A_n)e^{\alpha x}$
$P_n(x)e^{\alpha x}\left\{\begin{array}{c}\sin \omega x \\ \cos \omega x\end{array}\right\}$	$x^s[(A_0 x^n + A_1 x^{n-1} + \cdots + A_n)e^{\alpha x}\cos \omega x + (B_0 x^n + B_1 x^{n-1} + \cdots + B_n)e^{\alpha x}\sin \omega x]$

*$P_n(x)$ is a polynomial of degree n.

9. LAPLACE TRANSFORMS

$$\mathcal{L}\big(f(t)\big) = F(s) = \int_0^\infty e^{-st}f(t)dt \qquad 11.26$$

11. ALGEBRA OF LAPLACE TRANSFORMS

- *linearity theorem (c is a constant.)*

$$\mathcal{L}\big(cf(t)\big) = c\mathcal{L}\big(f(t)\big) = cF(s) \qquad 11.28$$

- *superposition theorem ($f(t)$ and $g(t)$ are different functions.)*

$$\mathcal{L}\big(f(t) \pm g(t)\big) = \mathcal{L}\big(f(t)\big) \pm \mathcal{L}\big(g(t)\big)$$
$$= F(s) \pm G(s) \qquad 11.29$$

- *time-shifting theorem (delay theorem)*

$$\mathcal{L}\big(f(t-b)u_b\big) = e^{-bs}F(s) \qquad 11.30$$

- *Laplace transform of a derivative*

$$\mathcal{L}\big(f^n(t)\big) = -f^{n-1}(0) - sf^{n-2}(0) - \cdots$$
$$- s^{n-1}f(0) + s^n F(s) \qquad 11.31$$

- *other properties*

$$\mathcal{L}\left(\int_0^t f(u)du\right) = \left(\frac{1}{s}\right)F(s) \qquad 11.32$$

$$\mathcal{L}\big(tf(t)\big) = -\frac{dF}{ds} \qquad 11.33$$

$$\mathcal{L}\left(\frac{1}{t}f(t)\right) = \int_s^\infty F(u)du \qquad 11.34$$

12. CONVOLUTION INTEGRAL

$$f(t) = \mathcal{L}^{-1}\big(F_1(s)F_2(s)\big)$$

$$= \int_0^t f_1(t - \chi)f_2(\chi)d\chi$$

$$= \int_0^t f_1(\chi)f_2(t - \chi)d\chi \qquad 11.35$$

EERM Chapter 12
Probability and Statistical Analysis of Data

> Chapter, section, equation, figure, and table numbers correspond to EERM. For additional study material, go to the corresponding chapter and section number in EERM.

2. COMBINATIONS OF ELEMENTS

- *binomial coefficient*

$$\binom{n}{r} = C(n,r) = \frac{n!}{(n-r)!r!} \quad \text{[for } r \le n] \qquad 12.19$$

3. PERMUTATIONS

$$P(n,r) = \frac{n!}{(n-r)!} \qquad \text{[for } r \le n] \qquad 12.20$$

9. BINOMIAL DISTRIBUTION

All outcomes can be categorized as either successes or failures.

$$p\{x\} = f(x) = \binom{n}{x}\hat{p}^x\hat{q}^{n-x} \qquad 12.28$$

$$\binom{n}{x} = \frac{n!}{(n-x)!x!} \qquad 12.29$$

- *mean,* μ

$$\mu = n\hat{p} \qquad 12.30$$

- *variance,* σ^2

$$\sigma^2 = n\hat{p}\hat{q} \qquad 12.31$$

12. POISSON DISTRIBUTION

Events occur relatively infrequently but at a relatively regular rate.

$$p\{x\} = f(x) = \frac{e^{-\lambda}\lambda^x}{x!} \quad [\lambda > 0] \qquad 12.34$$

λ is both the mean and the variance of the Poisson distribution.

15. NORMAL DISTRIBUTION

The *normal distribution (Gaussian distribution)* is a symmetrical distribution commonly referred to as the *bell-shaped curve*, with mean μ and variance σ^2.

$$f(x) = \frac{e^{-\frac{1}{2}\left(\frac{x-\mu}{\sigma}\right)^2}}{\sigma\sqrt{2\pi}} \quad [-\infty < x < +\infty] \qquad 12.41$$

$$p\{\mu < X < x_0\} = F(x_0)$$

$$= \frac{1}{\sigma\sqrt{2\pi}}\int_0^{x_0} e^{-\frac{1}{2}\left(\frac{x-\mu}{\sigma}\right)^2}dx \quad 12.42$$

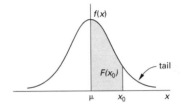

Figure 12.4 *Normal Distribution*

EERM Chapter 13
Computer Mathematics

> Chapter, section, equation, figure, and table numbers correspond to EERM. For additional study material, go to the corresponding chapter and section number in EERM.

1. POSITIONAL NUMBERING SYSTEMS

$$(a_n a_{n-1} \cdots a_2 a_1 a_0)_b = a_n b^n + a_{n-1}b^{n-1} + \cdots$$
$$+ a_2 b^2 + a_1 b + a_0 \qquad 13.1$$

2. CONVERTING BASE-*b* NUMBERS TO BASE-10

$$(0.a_1 a_2 \cdots a_m)_b = a_1 b^{-1} + a_2 b^{-2} + \cdots + a_m b^{-m} \qquad 13.2$$

4. BINARY NUMBER SYSTEM

- *two binary digits (bits)* (zero and one)

$$0 + 0 = 0$$
$$0 + 1 = 1$$
$$1 + 0 = 1$$
$$1 + 1 = 0 \text{ carry } 1$$

7. CONVERSIONS AMONG BINARY, OCTAL, AND HEXADECIMAL NUMBERS

Table 13.1 Binary, Octal, Decimal, and Hexadecimal Equivalents

binary	octal	decimal	hexadecimal
0	0	0	0
1	1	1	1
10	2	2	2
11	3	3	3
100	4	4	4
101	5	5	5
110	6	6	6
111	7	7	7
1000	10	8	8
1001	11	9	9
1010	12	10	A
1011	13	11	B
1100	14	12	C
1101	15	13	D
1110	16	14	E
1111	17	15	F
10000	20	16	10

8. COMPLEMENT OF A NUMBER

The *complement*, N^*, of a number, N, for a machine with a maximum number, n, of digits per integer number stored is given by

$$N_b^* = b^n - N \qquad \text{13.3}$$
$$N_{b-1}^* = N_b^* - 1 \qquad \text{13.4}$$

In base-2 arithmetic, the *two's* and *one's complements* are

$$N_2^* = 2^n - N \qquad \text{13.7}$$
$$N_1^* = N_2^* - 1 \qquad \text{13.8}$$

9. APPLICATION OF COMPLEMENTS TO COMPUTER ARITHMETIC

$$(N^*)^* = N \qquad \text{13.9}$$
$$M - N = M + N^* \qquad \text{13.10}$$

The binary one's complement is easily found by switching all of the ones and zeros to zeros and ones, respectively.

12. BOOLEAN ALGEBRA LAWS

- *special properties of 0 and 1*

$$0 + x = x \qquad 0 \cdot x = 0$$
$$1 + x = 1 \qquad 1 \cdot x = x$$

- *idempotence laws*

$$x + x = x \qquad x \cdot x = x$$

- *complementation laws*

$$x + \overline{x} = 1 \qquad x \cdot \overline{x} = 0$$

- *involution*

$$\overline{(\overline{x})} = x$$

- *commutative laws*

$$x + y = y + x \qquad x \cdot y = y \cdot x$$

- *associative laws*

$$x + (y + z) = (x + y) + z$$
$$x \cdot (y \cdot z) = (x \cdot y) \cdot z$$

- *distributive laws*

$$x \cdot (y + z) = (x \cdot y) + (x \cdot z)$$
$$x + (y \cdot z) = (x + y) \cdot (x + z)$$

- *absorption laws*

$$x + (x \cdot y) = x \qquad x \cdot (x + y) = x$$
$$x + (\overline{x} \cdot y) = x + y \qquad x \cdot (\overline{x} + y) = x \cdot y$$

13. BOOLEAN ALGEBRA THEOREMS

- *simplification theorems*

$$xy + x\overline{y} = x \qquad (x + y)(x + \overline{y}) = x$$
$$x + xy = x \qquad x(x + y) = x$$
$$(x + \overline{y})y = xy \qquad x\overline{y} + y = x + y$$

- *De Morgan's laws*

$$\overline{(x + y + z + \cdots)} = \overline{x}\,\overline{y}\,\overline{z}\cdots$$
$$\overline{(xyz\cdots)} = \overline{x} + \overline{y} + \overline{z}\cdots$$
$$\overline{[f(x_1, x_2, x_3, \cdots, x_n, 0, 1, +, \cdot)]}$$
$$= f(\overline{x_1}, \overline{x_2}, \overline{x_3}, \cdots, \overline{x_n}, 1, 0, \cdot, +)$$

EERM Chapter 15
Advanced Engineering Mathematics

> Chapter, section, equation, figure, and table numbers correspond to EERM. For additional study material, go to the corresponding chapter and section number in EERM.

6. LAPLACE TRANSFORM: INITIAL AND FINAL VALUE THEOREMS

- *initial value theorem*

$$f(0) = \lim_{s \to \infty} sF(s) \qquad \text{15.24}$$

- *final value theorem*

$$f(\infty) = \lim_{s \to 0} sF(s) \qquad \text{15.25}$$

9. SPECIAL FUNCTIONS

- *unit step function*

$$u_t = \begin{cases} 0 & 0 < t < k \\ 1 & t > k \end{cases} \qquad \text{15.37}$$

- *unit finite impulse function*

$$I(h, t - t_0) = H(t - t_0) - H\big(t - (t_0 + h)\big) \qquad \text{15.38}$$

$$I(h, t - t_0) = \begin{cases} 0 & t < t_0 \text{ and } t > (t_0 + h) \\ 1 & t_0 < t < (t_0 + h) \end{cases} \qquad \text{15.39}$$

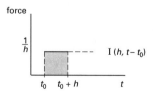

Figure 15.3 *Unit Finite Impulse Function*

- *delta function*

$$\lim_{h \to 0} I(h, t - t_0) \equiv \delta(t - t_0) = \begin{cases} 0 & t \neq t_0 \\ \infty & t = t_0 \end{cases} \qquad \text{15.40}$$

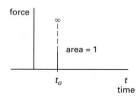

Figure 15.4 *Delta Function*

The delta function is strictly defined only in terms of integration.

$$\int_{-\infty}^{+\infty} \delta(t - t_0) f(t) dt = f(t_0) \qquad \text{15.41}$$

- *gamma function*

$$\Gamma(\alpha) = \int_{0}^{\infty} e^{-t} t^{(\alpha-1)} dt \quad [\alpha > 0] \qquad \text{15.42}$$

$$\Gamma(\alpha + 1) = \alpha \Gamma(\alpha) \qquad \text{15.43}$$

11. CONTINUOUS-TIME SYSTEMS: FOURIER SERIES REPRESENTATION OF A PERIODIC SIGNAL

- *Fourier series representation for any continuous-time periodic signal*

$$x(t) = \sum_{k=-\infty}^{+\infty} a_k e^{jk\omega_0 t} \qquad \text{15.58}$$

$$a_k = \frac{1}{T_0} \int_{T_0} x(t) e^{-jk\omega_0 t} dt \qquad \text{15.59}$$

12. CONTINUOUS-TIME SYSTEMS: FOURIER TRANSFORM OF AN APERIODIC SIGNAL

Consider the aperiodic signal as the limit of a periodic signal as the period becomes arbitrarily large.

- *Fourier transform* or *Fourier integral*

$$x(t) = \frac{1}{2\pi} \int_{-\infty}^{+\infty} X(\omega) e^{j\omega t} d\omega \qquad \text{15.60}$$

- *inverse Fourier transform* (that is, the *synthesis equation* of the pair)

$$X(\omega) = \int_{-\infty}^{+\infty} x(t) e^{-j\omega t} dt \qquad \text{15.61}$$

Equations 15.60 and 15.61 are referred to as a *Fourier transform pair*.

13. CONVOLUTION

- *convolution integral*

$$f(t) * g(t) = \int_{a}^{b} f(\tau) g(t - \tau) d\tau$$

$$= T^{-1}\big\{ F(s) G(s) \big\} \qquad \text{15.65}$$

14. LAPLACE TRANSFORMS

- *unilateral Laplace transform*

$$\mathcal{L}\big(f(t)\big) = F(s) = \int_0^\infty e^{-st} f(t)\,dt \qquad 15.70$$

- *bilateral Laplace transform*

$$\mathcal{L}\big(f(t)\big) = F(s) = \int_{-\infty}^{+\infty} e^{-st} f(t)\,dt \qquad 15.71$$

From the bilateral Laplace transform, the relationship between it and the Fourier transform is easily seen.

$$F(s)\big|_{s=j\omega} = \mathcal{F}\big(f(t)\big) \qquad 15.72$$

16. DISCRETE-TIME SYSTEMS: FOURIER SERIES REPRESENTATION OF A PERIODIC SIGNAL

$$k[n] = \sum_{k=\langle N \rangle} a_k e^{jk\left(\frac{2\pi}{N}\right)n} \qquad 15.75$$

$$a_k = \frac{1}{N} \sum_{n=\langle N \rangle} x[n] e^{-jk\left(\frac{2\pi}{N}\right)n} \qquad 15.76$$

17. DISCRETE-TIME SYSTEMS: FOURIER TRANSFORM OF AN APERIODIC SIGNAL

$$x[n] = \frac{1}{2\pi} \int_{2\pi} x[\Omega] e^{j\Omega n}\,d\Omega \qquad 15.77$$

- *discrete-time Fourier transform*

$$X[\Omega] = \sum_{n=-\infty}^{+\infty} x[n] e^{-j\Omega n} \qquad 15.78$$

Equations 15.77 and 15.78 are termed the *Fourier transform pair*. Equation 15.77 is the synthesis equation and Eq. 15.78 is the analysis equation.

18. *Z*-TRANSFORMS

The z-transform is the discrete-time counterpart to the Laplace transform. Also, it represents the generalization of the Fourier transform for discrete-time systems; that is, it is the transform to use when the value of s in a discrete-time system is $s = \sigma + j\Omega$.

$$X[z] = \sum_{n=-\infty}^{+\infty} x[n] z^{-n} \qquad 15.79$$

The relationship between the z-transform and the discrete-time Fourier transform is given by the following.

$$X[z]\big|_{z=e^{j\Omega}} = \mathcal{F}\big[x[n]\big] \qquad 15.80$$

19. TRANSFORMATION OF INTEGRALS

- *Gauss' divergence theorem*

$$\oiint_{A(V)} \mathbf{F}\cdot d\mathbf{A} = \iiint_{V(A)} \mathbf{\nabla}\cdot\mathbf{F}\,dV \qquad 15.82$$

- *Stokes' theorem*

$$\oint_{l(A)} \mathbf{F}\cdot d\mathbf{l} = \iint_{A(l)} \mathbf{\nabla}\times\mathbf{F}\cdot d\mathbf{A} \qquad 15.83$$

- *Green's theorem*

$$\oint_{l(A)} \big(f(x,y)dx + g(x,y)dy\big)$$

$$= \iint_{A(l)} \left(\frac{\delta g}{\delta x} - \frac{\delta f}{\delta y}\right) dx\,dy \qquad 15.84$$

21. COMPLEX NUMBERS

$$z = x + jy \qquad 15.93$$

j is used instead of i to represent $\sqrt{-1}$ to avoid confusion with the nomenclature for current.

$$\overline{z} = z^* = x - jy \qquad 15.94$$

$$\mathcal{R}e\,z = x = \frac{1}{2}(z + z^*) \qquad 15.95$$

$$\mathcal{I}m\,z = y = \frac{1}{2j}(z - z^*) \qquad 15.96$$

Theory

EERM Chapter 16
Electromagnetic Theory

Chapter, section, equation, figure, and table numbers correspond to EERM. For additional study material, go to the corresponding chapter and section number in EERM.

7. COULOMB'S LAW

$$\mathbf{F}_{1\text{-}2} = \left(\frac{Q_1 Q_2}{4\pi\epsilon r_{1\text{-}2}^2}\right)\mathbf{a}_{r_{1\text{-}2}} \qquad 16.3$$

$$\mathbf{F}_{1\text{-}2} = \mathbf{F}_2 = Q_2\mathbf{E}_1 \qquad 16.4$$

- *Gaussian SI form*

$$\mathbf{F}_{1\text{-}2} = \frac{kQ_1 Q_2}{\epsilon_r r^2}\mathbf{a}_{1\text{-}2} \qquad 16.5$$

ϵ_r represents the *relative permittivity*.

k has an approximate value of 8.987×10^9 N·m²/C².

8. ELECTRIC FIELDS

$$\mathbf{E} = \frac{Q}{4\pi\epsilon r^2}\mathbf{a} \qquad 16.6$$

$$\mathbf{E}_1 = \sum_{i=1}^{i}\frac{Q_i}{4\pi\epsilon^2 r_{i1}^2}\mathbf{a}_{r_{i1}} \qquad 16.7$$

$$\mathbf{E}_1 = \int_V \frac{\rho dV}{4\pi\epsilon r_{i1}^2}\mathbf{a}_{r_{i1}} \qquad 16.8$$

The total charge Q in the volume V that has the charge density ρ is

$$Q = \int_V \rho dV \qquad 16.9$$

$$E_{\text{uniform}} = \frac{V_{\text{plates}}}{r} \qquad 16.10$$

Table 16.6 Electric Fields and Capacitance for Various Configurations

isolated point charge	infinite coaxial capacitor
$E = \dfrac{Q}{4\pi\epsilon r^2}$ $C = 0$	$E = \dfrac{\rho_l}{2\pi\epsilon r}$ $(a < r < b)$ $\dfrac{C}{L} = \dfrac{2\pi\epsilon}{\ln\left(\frac{b}{a}\right)}$
isolated sphere	infinite line distribution inside an infinite cylinder
$E = \dfrac{Q}{4\pi\epsilon(r+R)^2}$ $(r > 0)$ $C = 4\pi\epsilon R$	$E = \dfrac{\rho_l}{2\pi\epsilon r}$
concentric spheres	infinite sheet distribution
$E = \dfrac{Q}{4\pi\epsilon r^2}$ $(a < r < b)$ $C = \dfrac{4\pi\epsilon ab}{b-a}$	$E = \dfrac{\rho_s}{2\epsilon}$ $\rho_s = \dfrac{Q}{A}$
infinite line distribution	infinite parallel plates
$E = \dfrac{\rho_l}{2\pi\epsilon r}$	$E = \dfrac{\rho_s}{\epsilon} = \dfrac{V}{r}$ $\dfrac{C}{A} = \dfrac{\epsilon}{r}$
infinite isolated capacitor	dipole (doublet)
$E = \dfrac{\rho_l}{2\pi\epsilon(r+R)}$ $(r > 0)$	$E = \dfrac{Qd}{4\pi\epsilon r^3}(2\cos\theta a_r + \sin\theta a_\theta)$

$$\epsilon = \epsilon_0 \epsilon_r \qquad 16.11$$

$$\epsilon_r = \frac{C_{\text{with dielectric}}}{C_{\text{vacuum}}} \qquad 16.12$$

$$\epsilon = \epsilon_0(1 + \chi_e) \qquad 16.13$$

χ_e represents the electric susceptibility.

11. ELECTRIC FLUX DENSITY

$$\mathbf{D} = \epsilon \mathbf{E} = \epsilon_0 \epsilon_r \mathbf{E} \qquad 16.15$$

$$D = |\mathbf{D}| = \frac{\psi}{A} = \frac{Q}{A} = \sigma \qquad 16.16$$

12. GAUSS' LAW FOR ELECTROSTATICS

$$\Psi = \oiint D\, dA = \oiint \sigma\, dA$$
$$= Q \quad [d\mathbf{A} \text{ parallel to } \mathbf{D}] \qquad 16.17$$

$$\Psi = \oiint \mathbf{D} \cdot d\mathbf{s} = \oiint D \cos\theta\, ds$$
$$= Q \quad [\text{arbitrary surface}] \qquad 16.18$$

$$\Psi = \iiint \rho\, dv = Q \quad [\text{arbitrary volume}] \qquad 16.19$$

13. CAPACITANCE AND ELASTANCE

$$Q = CV \qquad 16.22(a)$$

$$C = \frac{Q}{V} \qquad 16.22(b)$$

14. CAPACITORS

- *parallel plate capacitor*

$$C = \frac{\epsilon A}{r} \qquad 16.24$$

- *total energy, U (in J)*

$$U = \frac{1}{2}CV^2 = \frac{1}{2}VQ = \left(\frac{1}{2}\right)\left(\frac{Q^2}{C}\right) \qquad 16.25$$

For n capacitors in parallel,

$$C_{\text{total}} = C_1 + C_2 + C_3 + \cdots + C_n \qquad 16.26$$

For n capacitors in series,

$$\frac{1}{C_{\text{total}}} = \frac{1}{C_1} + \frac{1}{C_2} + \frac{1}{C_3} + \cdots + \frac{1}{C_n} \qquad 16.27$$

15. ENERGY DENSITY IN AN ELECTRIC FIELD

$$U_{\text{ave}} = \frac{1}{2}V_{\text{voltage}}Q = \left(\frac{1}{2}\right)(Et)(DA) = \frac{1}{2}EDV_{\text{volume}}$$

$$= \frac{1}{2}\epsilon E^2 V_{\text{volume}} = \left(\frac{1}{2}\right)\left(\frac{D^2}{\epsilon}\right)V_{\text{volume}} \qquad 16.28$$

Equation 16.28 in all its forms assumes that $\mathbf{D}$ and $\mathbf{E}$ are in the same direction.

$$u_{\text{ave}} = \frac{U_{\text{ave}}}{V_{\text{volume}}} = \frac{1}{2}\epsilon E^2 = \left(\frac{1}{2}\right)\left(\frac{D^2}{\epsilon}\right) \qquad 16.29$$

16. SPEED AND MOBILITY OF CHARGE CARRIERS

$$\mathbf{F} = Q\mathbf{E} = m\mathbf{a} \qquad 16.30$$

- *average drift velocity*

$$\mathbf{v}_d = \int \mathbf{a}\, dt = \int \frac{\mathbf{F}}{m}\, dt$$

$$= \frac{Q}{m}t_m\mathbf{E} = \mu\mathbf{E} \qquad 16.31$$

17. CURRENT

For simple DC circuits,

$$I = \frac{dQ}{dt} \qquad 16.33$$

18. CONVECTION CURRENT

$$I = \rho A \mathbf{v}_{\text{ave}} = \left(\frac{Q}{V}\right)A\mathbf{v}_{\text{ave}}$$

$$= \left(\frac{Q}{Al}\right)A\left(\frac{l}{t}\right) = \frac{Q}{t} \qquad 16.34$$

In terms of the current density vector,

$$\mathbf{J} = \rho\mathbf{v}_d \qquad 16.35$$

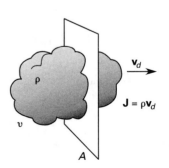

Figure 16.9 *Convection Current*

19. DISPLACEMENT CURRENT

$$i_d = \frac{d\psi}{dt} \qquad 16.36$$

$$\mathbf{J}_d = \frac{\delta \mathbf{D}}{\delta t} \qquad 16.37$$

$$i_d = \frac{d}{d_t} \int_A \mathbf{D} \cdot d\mathbf{A} \qquad 16.38$$

20. CONDUCTION CURRENT

$$\mathbf{J}_c = \rho \mathbf{v}_d \qquad 16.40$$

- *point form of Ohm's law*

$$\mathbf{J} = \sigma \mathbf{E} \qquad 16.41$$

21. MAGNETIC POLES

$$m = pd \qquad 16.42$$

22. BIOT-SAVART LAW

$$\mathbf{F}_{1\text{-}2} = \left(\frac{\mu}{4\pi}\right)\left(\frac{(I_2 d\mathbf{l}_2) \times (I_1 d\mathbf{l}_1) \times (\mathbf{r}_{1\text{-}2})}{r_{1\text{-}2}^2}\right) \qquad 16.43$$

For parallel currents,

$$F_{1\text{-}2} = \frac{\mu}{4\pi r^2} I_1 dl_1 I_2 dl_2 \qquad 16.44$$

- *magnetic force between two moving charges*

$$\mathbf{F}_{1\text{-}2} = \left(\frac{\mu}{4\pi}\right)\left(\frac{Q_1 Q_2}{r_{1\text{-}2}^2}\right)(\mathbf{v}_2)(\mathbf{v}_1 \times \mathbf{r}_{1\text{-}2}) \qquad 16.45$$

- *maximum magnetic force between parallel moving charges*

$$F = \frac{\mu}{4\pi r^2} Q_1 Q_2 \mathbf{v}^2 \qquad 16.46$$

- *maximum coulombic force between moving charges in free space*

$$F_e = \frac{Q_1 Q_2}{4\pi \epsilon_0 r^2} \qquad 16.47$$

- *maximum magnetic force between moving charges in free spaces*

$$F_m = \frac{\mu_0 Q_1 Q_2 \mathbf{v}^2}{4\pi r^2} \qquad 16.48$$

$$\frac{F_m}{F_e} = \epsilon_0 \mu_0 \mathbf{v}^2 \qquad 16.49$$

$$c^2 = \frac{1}{\mu_0 \epsilon_0} \qquad 16.50$$

23. MAGNETIC FIELDS

$$\mathbf{B} = \frac{\mu Il}{4\pi r^2} \mathbf{a} \qquad 16.52$$

$$B = |\mathbf{B}| = \frac{\phi}{A} \qquad 16.53$$

24. PERMEABILITY AND SUSCEPTIBILITY

$$\mu = \mu_0 \mu_r \qquad 16.55$$

$$\mu = \frac{B}{H} \qquad 16.56$$

$$\mu = \mu_0 (1 + \chi_m) \qquad 16.57$$

- *magnetic susceptibility, χ_m* (a dimensionless quantity)

$$\chi = \frac{M}{H} \qquad 16.58$$

25. MAGNETIC FLUX

$$\phi = \mu NIl = BA \qquad 16.60$$

The term NI is the *magnetomotive force*.

26. MAGNETIC FIELD STRENGTH

Table 16.8 *Magnetic Field and Inductance for Various Configurations*

straight infinite conductor

$$H = \frac{I}{2\pi r}$$

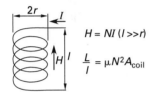
N loops

$$H = \frac{NI}{2r} \quad \text{(center of coil only)}$$

infinite cylindrical coil helix (solenoid)

$$H = NI \; (l >> r)$$

$$\frac{L}{l} = \mu N^2 A_{\text{coil}}$$

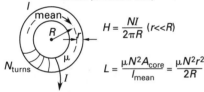
torus (toroidal coil)

$$H = \frac{NI}{2\pi R} \; (r << R)$$

$$L = \frac{\mu N^2 A_{\text{core}}}{l_{\text{mean}}} = \frac{\mu N^2 r^2}{2R}$$

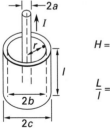
coaxial cable (high frequencies)

$$H = \frac{I}{2\pi r} \; (a < r < b)$$

$$\frac{L}{l} = \frac{\mu}{2\pi} \ln\left(\frac{b}{a}\right)$$

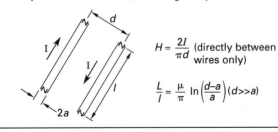
parallel transmission lines (high frequencies)

$$H = \frac{2I}{\pi d} \; \text{(directly between wires only)}$$

$$\frac{L}{l} = \frac{\mu}{\pi} \ln\left(\frac{d-a}{a}\right) (d >> a)$$

$$\mathbf{H} = \frac{1}{\mu}\mathbf{B} \qquad 16.61$$

28. INDUCTANCE AND RECIPROCAL INDUCTANCE

$$\mathbf{v} = L\frac{di}{dt} \qquad 16.63$$

29. INDUCTORS

For a solenoid,

$$L = \mu\frac{N^2 A_{\text{coil}}}{l} \qquad 16.67$$

- *total energy, U* (in J)

$$U = \frac{1}{2}LI^2 \qquad 16.68$$

For inductors in series,

$$L_{\text{total}} = L_1 + L_2 + L_3 + \ldots + L_n \qquad 16.69$$

For inductors in parallel,

$$\frac{1}{L_{\text{total}}} = \frac{1}{L_1} + \frac{1}{L_2} + \frac{1}{L_3} + \ldots + \frac{1}{L_n} \qquad 16.70$$

30. ENERGY DENSITY IN A MAGNETIC FIELD

$$U_{\text{ave}} = \frac{1}{2}\phi NI \qquad 16.71$$

$$u_{\text{ave}} = \frac{U_{\text{ave}}}{V_{\text{volume}}} = \frac{1}{2}BH = \frac{1}{2}\mu H^2 = \frac{1}{2}\frac{B^2}{\mu} \qquad 16.72$$

31. SPEED AND DIRECTION OF CHARGE CARRIERS

$$\mathbf{F} = Q\mathbf{v} \times \mathbf{B} \qquad 16.74$$

If the conductor is straight and the magnetic field is constant along the length of the conductor,

$$F = NIBl\sin\theta \qquad 16.76$$

If at right angles,

$$F = NIBl \qquad 16.77$$

- *Lorentz force equation*

$$\mathbf{F} = Q\left(\mathbf{E} + \mathbf{v} \times \mathbf{B}\right) \qquad 16.80$$

Electromagnetic waves propagate through space with a speed given by

$$v = \frac{1}{\sqrt{\epsilon\mu}} \qquad 16.81$$

32. VOLTAGE AND THE MAGNETIC CIRCUIT

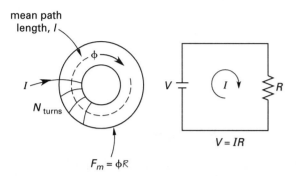

Figure 16.17 *Magnetic-Electric Circuit Analogy*

- *reluctance, $\mathcal{R}$*

$$\mathcal{R} = \frac{l}{\mu A} \qquad 16.84$$

The magnetic equation that correlates with Ohm's law in electric circuits is

$$F_m = Hl = \phi\mathcal{R} \qquad 16.85$$

33. MAGNETIC FIELD-INDUCED VOLTAGE

$$V = -N\frac{d\phi}{dt} \qquad 16.86$$

$$V = N\frac{d\phi}{dt} = NBl\frac{ds}{dt} = NBlv \qquad 16.87$$

EERM Chapter 17
Electronic Theory

Chapter, section, equation, figure, and table numbers correspond to EERM. For additional study material, go to the corresponding chapter and section number in EERM.

2. CHARGES IN A VACUUM

$$\mathbf{F}_{\parallel} = Q\mathbf{E} \qquad 17.1$$

$$\mathbf{F}_{\perp} = Q\mathbf{v} \times \mathbf{B} \qquad 17.2$$

The direction of motion of the particle will be a circle if the magnetic field is uniform and the particle enters it at right angles.

$$r = \frac{mv}{QB} \qquad 17.3$$

$$\mathbf{F} = \mathbf{F}_{\parallel} + \mathbf{F}_{\perp} = Q\left(\mathbf{E} + (\mathbf{v} \times \mathbf{B})\right) \qquad 17.4$$

- *photoelectric effect*

$$h\nu = \varphi + \frac{1}{2}mv^2 \qquad 17.6$$

3. CHARGES IN LIQUIDS AND GASES

- *conductivity and resistivity*

$$\sigma = \rho\mu \qquad 17.8$$

In metallic conductors, the conductivity is often defined by the point form of Ohm's law.

$$\sigma = \frac{|\mathbf{J}|}{|\mathbf{E}|} = \frac{J}{E} \qquad 17.9$$

4. CHARGES IN SEMICONDUCTORS

- *law of mass action*

$$np = n_i^2 \qquad 17.13$$

- *Einstein equation*

$$V_T = \frac{D_p}{\mu_p} = \frac{D_n}{\mu_n} \qquad 17.15$$

$$V_T = \frac{kT}{q} \qquad 17.16$$

EERM Chapter 18
Communication Theory

Chapter, section, equation, figure, and table numbers correspond to EERM. For additional study material, go to the corresponding chapter and section number in EERM.

1. FUNDAMENTALS

Analog and Digital Signals

The rate that achieves ideal sampling is called the *Nyquist rate*, f_S, and is given by

$$f_S = \frac{1}{T_S} = 2f_I \qquad 18.1$$

The *Nyquist interval*, T_S, is the sampling period corresponding to the Nyquist rate and is given by

$$T_S = \frac{1}{2f_I} \qquad 18.2$$

Basic Signal Theory

The primary uncertainty relationship is

$$\Delta T \Delta \Omega = 2\pi \qquad 18.4$$

The second moment uncertainty relationship, which relates the energy, is

$$\Delta T_2 \Delta \Omega_2 \geq \frac{1}{2} \qquad 18.5$$

Information Entropy

- *probability of occurrence*, $p(x_i)$
- *self-information*, $I(x_i)$

$$I(x_i) = \log\left(\frac{1}{p(x_i)}\right) = -\log p(x_i) \qquad 18.6$$

Decibels

$$r = 10 \log_{10}\left(\frac{P_2}{P_1}\right) \qquad 18.8$$

$$r = 10 \log\left(\frac{I_2}{I_1}\right)^2 = 10 \log\left(\frac{V_2}{V_1}\right)^2$$

$$= 20 \log\left(\frac{I_2}{I_1}\right) = 20 \log\left(\frac{V_2}{V_1}\right) \qquad 18.9$$

4. NOISE

- *thermal noise* (or *Johnson noise*)

$$\overline{v_n^2} = 4kTR(\text{BW}) \qquad 18.10$$

- *shot noise* (or *Schottky noise*)

$$\overline{i_n^2} = 2qI(\text{BW}) \qquad 18.13$$

EERM Chapter 19
Acoustic and Transducer Theory

Chapter, section, equation, figure, and table numbers correspond to EERM. For additional study material, go to the corresponding chapter and section number in EERM.

1. INTRODUCTION

$$v_s = f\lambda \qquad 19.1$$

$$T = \frac{1}{f} \qquad 19.2$$

2. SOUND WAVE PROPAGATION VELOCITY

The speed of sound in solids and liquids is given by

$$v_s = \sqrt{\frac{E}{\rho}} \qquad 19.3$$

For ideal gases, the propagation velocity is

$$v_s = \sqrt{\frac{\gamma R^* T}{\text{MW}}} \qquad 19.4$$

3. ENERGY AND INTENSITY OF SOUND WAVES

- *energy density*, η

$$\eta = \frac{1}{2}\frac{p_0^2}{\rho v^2} \qquad 19.5$$

- *intensity of the sound wave*

$$I = \eta v = \frac{1}{2}\frac{p_0^2}{\rho v} \qquad 19.6$$

- *sound-pressure level*, L_p

$$L_p = 20 \log\left(\frac{p}{p_0}\right) \qquad 19.8$$

Fields

EERM Chapter 20
Electrostatics

Chapter, section, equation, figure, and table numbers correspond to EERM. For additional study material, go to the corresponding chapter and section number in EERM.

Electrostatic Fields

$$\mathbf{F} = Q\mathbf{E} \qquad 20.2$$

1. POINT CHARGE

$$\mathbf{E} = \left(\frac{Q}{4\pi\epsilon r^2}\right)\mathbf{a}_r \qquad 20.4$$

$$V = \frac{Q}{4\pi\epsilon r} \qquad 20.5$$

2. DIPOLE CHARGE

$$\mathbf{p} = q\mathbf{d} \qquad 20.6$$

$$\mathbf{E} = \left(\frac{Qd}{4\pi\epsilon r^3}\right)(2\cos\theta\,\mathbf{a}_r + \sin\theta\,\mathbf{a}_\theta) \qquad 20.7$$

$$V = \frac{Q(R_2 - R_1)}{4\pi\epsilon R_1 R_2} \approx \frac{Qd\cos\theta}{4\pi\epsilon r^2} \qquad 20.8$$

3. COULOMB'S LAW

$$\mathbf{F} = \left(\frac{Q_1 Q_2}{4\pi\epsilon r^2}\right)\mathbf{a} \qquad 20.11$$

4. LINE CHARGE

$$\mathbf{E} = \left(\frac{\rho_l}{2\pi\epsilon r}\right)\mathbf{a}_r \qquad 20.12$$

5. PLANE CHARGE

$$\mathbf{E} = \left(\frac{\rho_s}{2\epsilon}\right)\mathbf{a}_n \qquad 20.14$$

Divergence

$$\text{div } \mathbf{D} = \nabla\cdot\mathbf{D} = \rho \qquad 20.18$$

Potential

The potential given in terms of the electric field is

$$\mathbf{E} = -\nabla V \qquad 20.19$$

The potential can be represented in integral form.

$$V_{\mathrm{BA}} = \frac{W}{\Delta Q} = -\int_{\mathrm{B}}^{\mathrm{A}} \mathbf{E}\cdot d\mathbf{l} \qquad 20.20$$

$$\text{div } \mathbf{E} = \nabla\cdot\mathbf{E} = \frac{\rho}{\epsilon} \qquad 20.21$$

$$\nabla^2 v = -\frac{\rho}{\epsilon} \qquad 20.22$$

Work and Energy

The total work done in moving a charge from point A to point B in an electric field is

$$W_{AB} = -\int_A^B Q\mathbf{E}\cdot d\mathbf{l}$$

$$= -\int_A^B QE\cos\theta\,dl \qquad 20.25$$

The energy "stored" in the volume of the electrostatic field is

$$U = \frac{1}{2}\int \rho V\,dv = \frac{1}{2}\int \mathbf{D}\cdot\mathbf{E}\,dv$$

$$= \frac{1}{2}\int \epsilon E^2\,dv = \frac{1}{2}\int \frac{D^2}{\epsilon}\,dv \qquad 20.26$$

$$W = -\oint Q\mathbf{E}\cdot d\mathbf{l} = 0 \qquad 20.27$$

EERM Chapter 21
Electrostatic Fields

> Chapter, section, equation, figure, and table numbers correspond to EERM. For additional study material, go to the corresponding chapter and section number in EERM.

1. POLARIZATION

$$\mathbf{P} = \chi_e \varepsilon_0 \mathbf{E} \qquad 21.3$$

$$\mathbf{D} = \varepsilon_0 \left(1 + \chi_e\right) \mathbf{E} \qquad 21.4$$

$$\mathbf{D} = \varepsilon \mathbf{E} = \varepsilon_0 \varepsilon_r \mathbf{E} = \varepsilon_0 \mathbf{E} + \mathbf{P} \qquad 21.7$$

3. ELECTRIC DISPLACEMENT

If the angle $\pi/2$ is chosen as the arbitrary zero energy reference, the work of rotating a dipole is

$$W = -\mathbf{p} \cdot \mathbf{E} \qquad 21.8$$

For a poor conductor, or *lossy dielectric*,

$$\frac{J_c}{J_d} = \frac{\sigma}{\omega \varepsilon} \qquad 21.10$$

EERM Chapter 22
Magnetostatics

> Chapter, section, equation, figure, and table numbers correspond to EERM. For additional study material, go to the corresponding chapter and section number in EERM.

The Magnetic Field

The strength of the **B**-field is

$$\mathbf{B} = \left(\frac{\mu I l}{4\pi r^2}\right) \mathbf{a} \qquad 22.1$$

The term **B** is most often called the magnetic flux density.

$$B = \frac{\Phi}{A} \qquad 22.2$$

1. BIOT-SAVART LAW

$$\mathbf{B} = \left(\frac{\mu_0}{4\pi}\right)\left(\frac{q\mathbf{v} \times \mathbf{r}}{r^2}\right) \qquad 22.7$$

2. FORCE ON A MOVING CHARGED PARTICLE

$$\mathbf{F} = Q\mathbf{v} \times \mathbf{B} \qquad 22.8$$

A charged particle in a uniform magnetic field will travel in a circular path with radius and angular velocity of

$$r = \frac{m\mathrm{v}}{QB} \qquad 22.9$$

$$\omega = \frac{QB}{m} \qquad 22.10$$

$$B = \frac{F}{q\mathrm{v}} \qquad 22.11$$

3. FORCE ON CURRENT ELEMENTS

In terms of magnitude and assuming a uniform magnetic field,

$$F = lIB \sin\theta \qquad 22.15$$

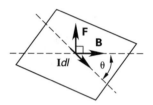

(a) magnetic force direction in terms of I*dl*

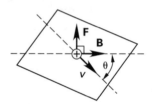

(b) magnetic force direction in terms of individual charge velocities

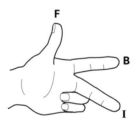

(c) FBI rule

Figure 22.2 *Magnetic Force Direction*

For a straight, infinitely long conductor,

$$B = \frac{\mu I}{2\pi r} \qquad 22.16$$

For two straight, infinitely long conductors,

$$\frac{\mathbf{F}}{l} = \mathbf{I} \times \mathbf{B} = \left(\frac{\mu I_1 I_2}{2\pi r}\right)\mathbf{a} \qquad 22.17$$

Lorentz Force Law

$$\mathbf{F} = \mathbf{F}_e + \mathbf{F}_m = Q\left(\mathbf{E} + \mathbf{v} \times \mathbf{B}\right) \qquad 22.19$$

Traditional Magnetism

Table 22.1 *Magnetic Units*

quantity	symbol	cgs units	SI units	conversion: cgs to SI
pole strength or flux	p or Φ or ϕ	maxwells	Wb	10^8 maxwells $= 1$ Wb
flux density	B	gauss	T	10^4 gauss $= 1$ T
field strength (intensity)	H	oersted	A/m	$4\pi \times 10^{-3}$ oersted $= 1$ A/m
magnetization[a]	M	oersted	A/m	$4\pi \times 10^{-3}$ oersted $= 1$ A/M
permeability	μ	gauss/oersted	H/m	$\mu_{cgs}\left(4\pi \times 10^{-7}\right) = \mu_{mks}$[b]

[a]The units used in this row assume the following equation for flux density: $B = \mu_0 H + \mu_0 M$. Other forms of this equation exist. Care must be used in determining the correct units or in comparing values from different references.
[b]The value of μ_{cgs} in the cgs system is 1.

6. MAGNETIC DIPOLE

- *magnetic moment,* **m**

$$\mathbf{m} = IA\mathbf{n} \qquad 22.27$$

$$\mathbf{T} = \mathbf{m} \times \mathbf{B} \qquad 22.28$$

7. COULOMB'S LAW EQUIVALENT: FORCE BETWEEN MAGNETIC POLES

$$\mathbf{F}_{1-2} = p_2\mathbf{H}_1 = \left(\frac{p_1 p_2}{4\pi\mu r^2}\right)\mathbf{a} \qquad 22.30$$

Curl

In magnetostatics, no magnetic charge exists.

$$\nabla\cdot\mathbf{B} = 0 \qquad 22.38$$

The curl of the **B**-field in free space is

$$\nabla \times \mathbf{B} = \mu_0\mathbf{J} \qquad 22.39$$

In terms of the magnetic field strength,

$$\nabla \times \mathbf{H} = \mathbf{J} \qquad 22.40$$

EERM Chapter 23
Magnetostatic Fields

Chapter, section, equation, figure, and table numbers correspond to EERM. For additional study material, go to the corresponding chapter and section number in EERM.

1. MAGNETIZATION

$$\mathbf{B} = \mu\mathbf{H} = \mu_0\mu_r\mathbf{H} = \mu_0\mathbf{H} + \mu_0\mathbf{M} \qquad 23.8$$

3. AUXILIARY FIELD H

$$H = \frac{NI}{l} \qquad 23.10$$

EERM Chapter 24
Electrodynamics

Chapter, section, equation, figure, and table numbers correspond to EERM. For additional study material, go to the corresponding chapter and section number in EERM.

1. ELECTROMOTIVE FORCE

$$\mathcal{E} = \oint \mathbf{F}\cdot d\mathbf{l} = \oint \frac{\mathbf{J}}{\sigma}\cdot d\mathbf{l}$$

$$= \oint \frac{I}{A\sigma}dl = I\oint \frac{1}{A\sigma}dl = IR \qquad 24.1$$

Electromotive force can be generated by time-varying flux.

$$v = -N\frac{d\phi}{dt} \qquad 24.2$$

$$v = -\frac{d\phi}{di}\frac{di}{dt} = -L\frac{di}{dt} \qquad 24.3$$

Electromotive force can be generated by the flux-cutting method.

$$V = -NBl\mathbf{v} \qquad 24.4$$

3. FARADAY'S LAW OF ELECTROLYSIS

- *Faraday's law of electrolysis*

$$m_g = \frac{ItW_e}{1F} \qquad 24.8$$

The term W_e is the equivalent weight of an element, which is the atomic weight in grams divided by the valence change that occurs during electrolysis, z.

$$W_e = \frac{\text{AW}}{z} \qquad 24.9$$

5. ENERGY AND MOMENTUM: POYNTING'S VECTOR

$$\mathbf{S} = c\varepsilon E^2 \mathbf{a} = c\mu H^2 \mathbf{a} = \mathbf{E} \times \mathbf{H} \qquad 24.12$$

$$\mathcal{P}_{\text{ave}} = \tfrac{1}{2}(\text{Re})\,(\mathbf{E} \times \mathbf{H}^*) \qquad 24.13$$

EERM Chapter 25
Maxwell's Equations

Chapter, section, equation, figure, and table numbers correspond to EERM. For additional study material, go to the corresponding chapter and section number in EERM.

Table 25.3 *Electromagnetic Field Vector Equations*

$$\mathbf{D} = \varepsilon\mathbf{E} = \varepsilon_0\mathbf{E} + \mathbf{P} = \varepsilon_0\left(1 + \chi_e\right)\mathbf{E}$$
$$\mathbf{B} = \mu\mathbf{H} = \mu_0\mathbf{H} + \mu_0\mathbf{M} = \mu_0\left(1 + \chi_m\right)\mathbf{H}$$
$$\mathbf{J} = \sigma\mathbf{E} = \rho\mathbf{v}$$

3. COMPARISON OF ELECTRIC AND MAGNETIC EQUATIONS

Table 25.4 *Electric and Magnetic Circuit Analogies*

electric	magnetic
$\text{emf} = V = IR$	$\text{mmf} = V_m = \phi\mathcal{R}$
current I	flux ϕ
emf $\mathcal{E}$ or V	mmf V_m
resistance $R = \rho l/A = l/\sigma A$	reluctance $\mathcal{R} = l/\mu A$
resistivity ρ	reluctivity $1/\mu$
conductance $G = 1/R$	permeance $P_m = \mu A/l$
conductivity $\sigma = 1/\rho$	permeability μ

Circuit Theory

EERM Chapter 26
DC Circuit Fundamentals

Chapter, section, equation, figure, and table numbers correspond to EERM. For additional study material, go to the corresponding chapter and section number in EERM.

3. RESISTANCE

$$R = \frac{\rho l}{A} \qquad 26.1$$

$$A_{\text{cmil}} = \left(\frac{d_{\text{inches}}}{0.001}\right)^2 \qquad 26.2$$

- *thermal coefficient of resistivity,* α

$$R = R_0\left(1 + \alpha\Delta T\right) \qquad 26.5$$
$$\rho = \rho_0\left(1 + \alpha\Delta T\right) \qquad 26.6$$

4. CONDUCTANCE

$$\sigma = \frac{1}{\rho} \qquad 26.7$$

$$\% \text{ conductivity} = \frac{\sigma}{\sigma_{\text{Cu}}} \times 100\%$$
$$= \frac{\rho_{\text{Cu}}}{\rho} \times 100\% \qquad 26.9$$

$$\rho_{\text{Cu},20°\text{C}} = 1.7241 \times 10^{-6} \ \Omega\cdot\text{cm}$$
$$= 0.3403 \ \Omega\cdot\text{cmil/cm} \qquad 26.10$$

5. OHM'S LAW

$$V = IR \qquad 26.11$$

6. POWER

$$P = \frac{W}{t} = \frac{VIt}{t} = VI = \frac{V^2}{R} \qquad 26.12$$
$$P = I^2 R \qquad 26.13$$

8. ENERGY SOURCES

$$\text{regulation} = \frac{V_{\text{nl}} - V_{\text{fl}}}{V_{\text{fl}}} \times 100\% \qquad 26.15$$

13. VOLTAGE AND CURRENT DIVIDERS

$$V_2 = V_s\left(\frac{R_2}{R_1 + R_2}\right) \qquad 26.21$$

$$I_2 = I_s\left(\frac{R_1}{R_1 + R_2}\right) = I_s\left(\frac{G_2}{G_1 + G_2}\right) \qquad 26.23$$

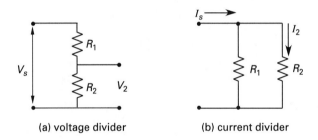

(a) voltage divider   (b) current divider

Figure 26.5 *Divider Circuits*

15. KIRCHHOFF'S VOLTAGE LAW

$$\sum_{\text{loop}} \text{voltage rises} = \sum_{\text{loop}} \text{voltage drops} \qquad 26.24$$

16. KIRCHHOFF'S CURRENT LAW

$$\sum_{\text{node}} \text{currents in} = \sum_{\text{node}} \text{currents out} \qquad 26.25$$

17. SERIES CIRCUITS

$$I = IR_1 = IR_2 = IR_3 \cdots = IR_N \qquad 26.26$$
$$R_e = R_1 + R_2 + R_3 \cdots + R_N \qquad 26.27$$
$$V_e = IR_e \qquad 26.29$$

18. PARALLEL CIRCUITS

$$V = V_1 = V_2 = V_3 \cdots = V_N$$

$$\frac{1}{R_e} = \frac{1}{R_1} + \frac{1}{R_2} + \frac{1}{R_3} \cdots + \frac{1}{R_N} \qquad 26.31(a)$$

$$I = \frac{V}{R_1} + \frac{V}{R_2} + \frac{V}{R_3} \cdots + \frac{V}{R_N} \qquad 26.32$$

20. DELTA-WYE TRANSFORMATIONS

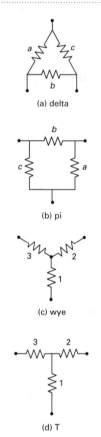

(a) delta

(b) pi

(c) wye

(d) T

Figure 26.10 *Delta (Pi)-Wye (T) Configurations*

$$R_a = \frac{R_1 R_2 + R_1 R_3 + R_2 R_3}{R_3} \qquad 26.33$$

$$R_b = \frac{R_1 R_2 + R_1 R_3 + R_2 R_3}{R_1} \qquad 26.34$$

$$R_c = \frac{R_1 R_2 + R_1 R_3 + R_2 R_3}{R_2} \qquad 26.35$$

$$R_1 = \frac{R_a R_c}{R_a + R_b + R_c} \qquad 26.36$$

$$R_2 = \frac{R_a R_b}{R_a + R_b + R_c} \qquad 26.37$$

$$R_3 = \frac{R_b R_c}{R_a + R_b + R_c} \qquad 26.38$$

EERM Chapter 27
AC Circuit Fundamentals

> Chapter, section, equation, figure, and table numbers correspond to EERM. For additional study material, go to the corresponding chapter and section number in EERM.

2. VOLTAGE

$$\omega = 2\pi f = \frac{2\pi}{T} \qquad 27.3$$

- *trigonometric*: $V_m \sin(\omega t + \theta)$
- *exponential*: $V_m e^{j\theta}$
- *polar or phasor*: $V_m \angle \theta$
- *rectangular*: $V_r + jV_i$

4. IMPEDANCE

$$\mathbf{Z} \equiv R \pm jX \qquad 27.6$$

$$R = Z \cos \phi \qquad \begin{bmatrix} \text{resistive or} \\ \text{real part} \end{bmatrix} \qquad 27.7$$

$$X = Z \sin \phi \qquad \begin{bmatrix} \text{reactive or} \\ \text{imaginary part} \end{bmatrix} \qquad 27.8$$

Table 27.1 *Characteristics of Resistors, Capacitors, and Inductors*

value	resistor R (Ω)	capacitor C (F)	inductor L (H)
reactance, X	0	$\dfrac{1}{\omega C}$	ωL
rectangular impedance, $\mathbf{Z}$	$R + j0$	$0 - \dfrac{j}{\omega C}$	$0 + j\omega L$
phasor impedance, $\mathbf{Z}$	$R\angle 0°$	$\dfrac{1}{\omega C}\angle -90°$	$\omega L\angle 90°$
phase	in-phase	leading	lagging
rectangular admittance, $\mathbf{Y}$	$\dfrac{1}{R} + j0$	$0 + j\omega C$	$0 - \dfrac{j}{\omega L}$
phasor admittance, $\mathbf{Y}$	$\dfrac{1}{R}\angle 0°$	$\omega C\angle 90°$	$\dfrac{1}{\omega L}\angle -90°$

7. AVERAGE VALUE

$$f_{\text{ave}} = \frac{1}{T} \int_{t_1}^{t_1+T} f(t)dt \qquad 27.14$$

$$f_{\text{ave}} = \frac{\text{positive area} - \text{negative area}}{T} \qquad 27.15$$

$$V_{\text{ave}} = \frac{1}{\pi} \int_0^{\pi} v(\theta)d\theta$$

$$= \frac{2V_m}{\pi} \quad \text{[rectified sinusoid]} \qquad 27.17$$

8. ROOT-MEAN-SQUARE VALUE

$$f_{\text{rms}}^2 = \frac{1}{T} \int_{t_1}^{t_1+T} f^2(t)dt \qquad 27.18$$

- *form factor*

$$\text{FF} = \frac{V_{\text{eff}}}{V_{\text{ave}}} \qquad 27.20$$

- *crest factor*, CF (also known as *peak factor* or *amplitude factor*)

$$\text{CF} = \frac{V_m}{V_{\text{eff}}} \qquad 27.21$$

9. PHASE ANGLES

A leading circuit is termed a *capacitive circuit*. A lagging circuit is termed an *inductive circuit*.

$$v(t) = V_m \sin(\omega t + \theta) \quad \text{[reference]} \qquad 27.24$$

$$i(t) = I_m \sin(\omega t + \theta + \phi) \quad \text{[leading]} \qquad 27.25$$

$$i(t) = I_m \sin(\omega t + \theta - \phi) \quad \text{[lagging]} \qquad 27.26$$

12. COMPLEX REPRESENTATION

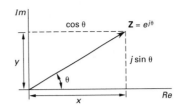

Figure 27.9 *Complex Quantities*

17. OHM'S LAW

$$\mathbf{V} = \mathbf{I}\mathbf{Z} \qquad 27.41$$

18. POWER

- *average power*

$$P_R = \tfrac{1}{2}I_m V_m = \frac{V_m^2}{2R} = I_{\text{rms}}V_{\text{rms}} = IV \qquad 27.44$$

19. REAL POWER AND THE POWER FACTOR

$$P = IV\cos\Psi = IV\,\text{pf} \qquad 27.53$$

$$P = \text{Re}\{\mathbf{V}\mathbf{I}^*\} \qquad 27.55$$

20. REACTIVE POWER

$$Q = IV\sin\Psi \qquad 27.56$$

Table 27.3 *Properties of Complex Numbers*

	rectangular form	polar/exponential form
	$\mathbf{Z} = x + jy$	$\mathbf{Z} = \lvert\mathbf{Z}\rvert\angle\theta$ $\mathbf{Z} = \lvert\mathbf{Z}\rvert e^{j\theta} = \lvert\mathbf{Z}\rvert\cos\theta + j\lvert\mathbf{Z}\rvert\sin\theta$
relationship between forms	$x = \lvert\mathbf{Z}\rvert\cos\theta$ $y = \lvert\mathbf{Z}\rvert\sin\theta$	$\lvert\mathbf{Z}\rvert = \sqrt{x^2 + y^2}$ $\theta = \tan^{-1}\left(\dfrac{y}{x}\right)$
complex conjugate	$\mathbf{Z}^* = x - jy$ $\mathbf{Z}\mathbf{Z}^* = \left(x^2 + y^2\right) = \lvert z\rvert^2$	$\mathbf{Z}^* = \lvert\mathbf{Z}\rvert e^{-j\theta} = \lvert\mathbf{Z}\rvert\angle-\theta$ $\mathbf{Z}\mathbf{Z}^* = (\lvert\mathbf{Z}\rvert e^{j\theta})(\lvert\mathbf{Z}\rvert e^{-j\theta}) = \lvert\mathbf{Z}\rvert^2$
addition	$\mathbf{Z}_1 + \mathbf{Z}_2 = (x_1 + x_2) + j(y_1 + y_2)$	$\mathbf{Z}_1 + \mathbf{Z}_2 = (\lvert\mathbf{Z}_1\rvert\cos\theta_1 + \lvert\mathbf{Z}_2\rvert\cos\theta_2)$ $+ j(\lvert\mathbf{Z}_1\rvert\sin\theta_1 + \lvert\mathbf{Z}_2\rvert\sin\theta_2)$
multiplication	$\mathbf{Z}_1\mathbf{Z}_2 = (x_1 x_2 - y_1 y_2) + j(x_1 y_2 + x_2 y_1)$	$\mathbf{Z}_1\mathbf{Z}_2 = \lvert\mathbf{Z}_1\rvert\lvert\mathbf{Z}_2\rvert\angle\theta_1 + \theta_2$
division	$\dfrac{\mathbf{Z}_1}{\mathbf{Z}_2} = \dfrac{(x_1 x_2 + y_1 y_2) + j(x_2 y_1 - x_1 y_2)}{\lvert\mathbf{Z}_2\rvert^2}$	$\dfrac{z_1}{z_2} = \dfrac{\lvert\mathbf{Z}_1\rvert}{\lvert\mathbf{Z}_2\rvert}\angle\theta_1 - \theta_2$

21. APPARENT POWER

$$S = IV \qquad 27.57$$

22. COMPLEX POWER: THE POWER TRIANGLE

$$S^2 = P^2 + Q^2 \qquad 27.58$$

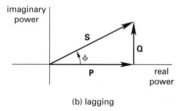

(a) leading

(b) lagging

Figure 27.11 *Power Triangle*

23. MAXIMUM POWER TRANSFER

The conditions for maximum power transfer, and thus resonance, are given by

$$R_{\text{load}} = R_s \qquad 27.61$$

$$X_{\text{load}} = -X_s \qquad 27.62$$

EERM Chapter 28
Transformers

Chapter, section, equation, figure, and table numbers correspond to EERM. For additional study material, go to the corresponding chapter and section number in EERM.

2. MAGNETIC COUPLING

$$V_s(t) = \frac{-N_s \left(2\pi f\right)\Phi_m \cos\omega t}{\sqrt{2}}$$

$$= -4.44\, N_s f \Phi_m \cos\omega t \qquad 28.3$$

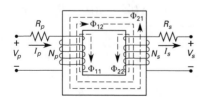

Figure 28.1 *Fluxes in a Magnetically Coupled Circuit*

$$V_p = I_p \left(R_p + j\omega L_p\right) - I_s j\omega M$$

$$= I_p \left(R_p + jX_p\right) - I_s jX_m$$

$$= I_p Z_p - I_s jX_m \qquad 28.7$$

$$V_s = I_s \left(R_s + j\omega L_s\right) - I_p j\omega M$$

$$= I_s \left(R_s + jX_s\right) - I_p jX_m$$

$$= I_s Z_s - I_p jX_m \qquad 28.8$$

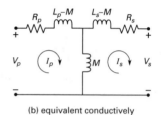

(a) flux linked circuit with mutual inductance

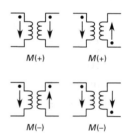

(b) equivalent conductively coupled circuit

Figure 28.2 *Transformer Models*

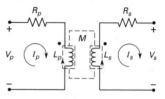

$M(+)$ $M(+)$

$M(-)$ $M(-)$

Figure 28.4 *Positive and Negative Mutual Inductances*

- *coupling coefficient (coefficient of coupling), k*

$$k = \frac{M}{\sqrt{L_p L_s}} = \frac{X_m}{\sqrt{X_p X_s}} \qquad 28.9$$

3. IDEAL TRANSFORMERS

$$a = \frac{N_p}{N_s} \qquad 28.10$$

$$I_p V_p = I_s V_s \qquad 28.11$$

$$a = \frac{N_p}{N_s} = \frac{V_p}{V_s} = \frac{I_s}{I_p} = \sqrt{\frac{Z_p}{Z_s}} \qquad 28.12$$

4. IMPEDANCE MATCHING

Z_{ep} is termed the *effective primary impedance* or the *reflected impedance*, Z_{ref}.

$$Z_{\mathrm{ep}} = Z_{\mathrm{ref}} = \frac{V_p}{I_p} = Z_p + a^2 Z_s \qquad 28.14$$

$$a = \sqrt{\frac{Z_p}{Z_s}} \qquad 28.15$$

$$R_p = a^2 R_s \qquad 28.16$$

$$X_p = -a^2 X_s \qquad 28.17$$

5. REAL TRANSFORMERS

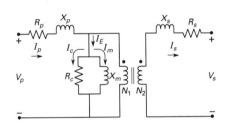

Figure 28.8 *Real Transformer Equivalent Circuit*

$$X_p = \omega L_p = \frac{V_{X_p}}{I_p} = \frac{4.44 f \Phi_p N_p}{I_p} \qquad 28.18$$

$$X_s = \omega L_s = \frac{V_{X_s}}{I_s} = \frac{4.44 f \Phi_s N_s}{I_s} \qquad 28.19$$

$$a = \sqrt{\frac{L_p}{L_s}} \qquad 28.20$$

6. MAGNETIC HYSTERESIS: BH CURVES

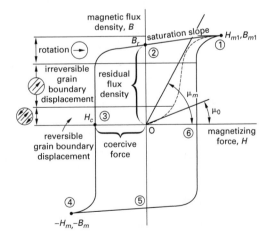

Figure 28.9 *Magnetic Hysteresis Loop*

EERM Chapter 29
Linear Circuit Analysis

> Chapter, section, equation, figure, and table numbers correspond to EERM. For additional study material, go to the corresponding chapter and section number in EERM.

9. RESISTANCE

For copper, α_{20} is approximately 3.9×10^{-3} C^{-1}.

$$\rho = \rho_{20} \left(1 + \alpha_{20} \left(T - 20°\mathrm{C} \right) \right) \qquad 29.3$$

The resistance of a conductor of length l and cross-sectional area A is

$$R = \frac{\rho l}{A} \qquad 29.4$$

- *Ohm's law for AC circuits*

$$V = IR \quad \text{or} \quad \mathbf{V} = \mathbf{I} Z_R \qquad 29.5$$

- *power dissipated*

$$P = IV = I^2 R = \frac{V^2}{R} \qquad 29.6$$

- *equivalent resistance of resistors in series*

$$R_e = R_1 + R_2 + \ldots R_n \qquad 29.7$$

- *equivalent resistance of resistors in parallel*

$$\frac{1}{R_e} = \frac{1}{R_1} + \frac{1}{R_2} + \ldots \frac{1}{R_n} \qquad 29.8$$

10. CAPACITANCE

Capacitance is a measure of the ability to store charge.

$$i = C \left(\frac{dv}{dt} \right) \qquad 29.9$$

For free space, that is, a vacuum, the permittivity is 8.854×10^{-12} F/m.

$$\varepsilon = \varepsilon_r \varepsilon_0 \qquad 29.10$$

- *simple parallel plate capacitor*

$$C = \frac{\varepsilon A}{r} \qquad 29.11$$

- *defining equation for capacitance*

$$C = \frac{Q}{V} \qquad 29.12$$

- *(average) energy stored*

$$U = \tfrac{1}{2}CV^2 = \tfrac{1}{2}QV = \tfrac{1}{2}\left(\frac{Q^2}{C}\right) \qquad 29.13$$

- *equivalent capacitance of capacitors in series*

$$\frac{1}{C_e} = \frac{1}{C_1} + \frac{1}{C_2} + \ldots + \frac{1}{C_n} \qquad 29.14$$

- *equivalent capacitance of capacitors in parallel*

$$C_e = C_1 + C_2 + \ldots + C_n \qquad 29.15$$

11. INDUCTANCE

Inductance is a measure of the ability to store magnetic energy.

$$v = L\left(\frac{di}{dt}\right) \qquad 29.16$$

For free space, that is, a vacuum, the permeability is 1.2566×10^{-6} H/m.

$$\mu = \mu_r \mu_0 \qquad 29.17$$

For a simple toroid,

$$L = \frac{\mu N^2 A}{l} \qquad 29.18$$

- *(average) energy stored*

$$U = \tfrac{1}{2}LI^2 = \tfrac{1}{2}\Psi I = \tfrac{1}{2}\left(\frac{\Psi^2}{L}\right) \qquad 29.20$$

- *equivalent inductance of inductors in series*

$$L_e = L_1 + L_2 + \ldots + L_n \qquad 29.21$$

- *equivalent inductance of inductors in parallel*

$$\frac{1}{L_e} = \frac{1}{L_1} + \frac{1}{L_2} + \ldots + \frac{1}{L_n} \qquad 29.22$$

Table 29.1 *Linear Circuit Element Parameters*

circuit element	voltage	current	instantaneous power	average power	average energy stored
(resistor)	$v = iR$	$i = \dfrac{v}{R}$	$p = iv = i^2R = \dfrac{v^2}{R}$	$P = IV = I^2R = \dfrac{V^2}{R}$	–
(capacitor)	$v = \dfrac{1}{C}\displaystyle\int i\,dt + \kappa$	$i = C\left(\dfrac{dv}{dt}\right)$	$p = iv = Cv\left(\dfrac{dv}{dt}\right)$	0	$U = \tfrac{1}{2}CV^2$
(inductor)	$v = L\left(\dfrac{di}{dt}\right)$	$i = \dfrac{1}{L}\displaystyle\int v\,dt + \kappa$	$p = iv = Li\left(\dfrac{di}{dt}\right)$	0	$U = \tfrac{1}{2}LI^2$

12. MUTUAL INDUCTANCE

$$U = \tfrac{1}{2}L_1 I_1^2 + \tfrac{1}{2}L_2 I_2^2 + M I_1 I_2 \qquad 29.23$$

Table 29.2 *Linear Circuit Parameters, Time and Frequency Domain Representation*

parameter	defining equation	time domain	frequency domain[a]
resistance	$R = \dfrac{v}{i}$	$v = iR$	$V = IR$
capacitance	$C = \dfrac{Q}{V}$	$i = C\left(\dfrac{dv}{dt}\right)$	$I = j\omega C V$
self-inductance	$L = \dfrac{\Psi}{I}$	$v = L\left(\dfrac{di}{dt}\right)$	$V = j\omega L I$
mutual inductance[b]	$M_{12} = M_{21} = M = \dfrac{\Psi_{12}}{I_2}$ $= \dfrac{\Psi_{21}}{I_1}$	$v_1 = L_1\left(\dfrac{di_1}{dt}\right) + M\left(\dfrac{di_2}{dt}\right)$ $v_2 = L_2\left(\dfrac{di_2}{dt}\right) + M\left(\dfrac{di_1}{dt}\right)$	$V_1 = j\omega(L_1 I_1 + M I_2)$ $V_2 = j\omega(L_2 I_2 + M I_1)$

[a]The voltages and currents in the frequency domain column are not shown as vectors; for example, **I** or **V**. They are, however, shown in phasor form. Across any single circuit element (or parameter), the phase angle is determined by the impedance and embodied by the j. If the $j\omega$ (C or L or M) were not shown and instead given as **Z**, the current and voltage would be shown as **I** and **V**, respectively.
[b]Currents I_1 and I_2 (in the mutual inductance row) are in phase. The angle between either I_1 or I_2 and V_1 or V_2 is determined by the impedance angle embodied by the j. Even in a real transformer, this relationship holds, since the equivalent circuit accounts for any phase difference with a magnetizing current, I_m.

17. DELTA-WYE TRANSFORMATIONS

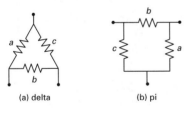

(a) delta   (b) pi

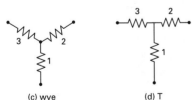

(c) wye   (d) T

Figure 29.10 *Delta (Pi)-Wye (T) Configurations*

$$Z_a = \frac{Z_1 Z_2 + Z_1 Z_3 + Z_2 Z_3}{Z_3} \qquad 29.26$$

$$Z_b = \frac{Z_1 Z_2 + Z_1 Z_3 + Z_2 Z_3}{Z_1} \qquad 29.27$$

$$Z_c = \frac{Z_1 Z_2 + Z_1 Z_3 + Z_2 Z_3}{Z_2} \qquad 29.28$$

$$Z_1 = \frac{Z_a Z_c}{Z_a + Z_b + Z_c} \qquad 29.29$$

$$Z_2 = \frac{Z_a Z_b}{Z_a + Z_b + Z_c} \qquad 29.30$$

$$Z_3 = \frac{Z_b Z_c}{Z_a + Z_b + Z_c} \qquad 29.31$$

18. THEVENIN'S THEOREM

$$\mathbf{V}_{Th} = \mathbf{V}_{oc} \qquad 29.32$$

$$\mathbf{Z}_{Th} = \frac{\mathbf{V}_{oc}}{\mathbf{I}_{sc}} \qquad 29.33$$

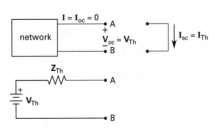

Figure 29.11 *Thevenin Equivalent Circuit*

19. NORTON'S THEOREM

$$\mathbf{I}_N = \mathbf{I}_{sc} \qquad 29.34$$

$$\mathbf{Z}_N = \frac{\mathbf{V}_{oc}}{\mathbf{I}_{sc}} \qquad 29.35$$

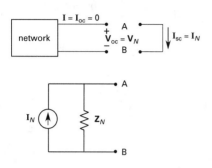

Figure 29.12 *Norton Equivalent Circuit*

20. MAXIMUM POWER TRANSFER THEOREM

Where the load impedance varies and the source impedance is fixed, maximum power transfer occurs when the load and source impedances are complex conjugates. That is, $\mathbf{Z}_l = \mathbf{Z}_s^*$ or $R_{\text{load}} + jX_{\text{load}} = R_s - jX_s$.

21. SUPERPOSITION THEOREM

step 1: Replace all sources except one by their internal resistances. Ideal current sources are replaced by open circuits. Ideal voltage sources are replaced by short circuits.

step 2: Compute the desired quantity, either voltage or current, for the element in question due to the single source.

step 3: Repeat steps 1 and 2 for each of the sources in turn.

step 4: Sum the calculated values obtained for the current or voltage obtained in step 2. The result is the actual value of the current or voltage in the element for the complete circuit.

Superposition is not valid for circuits in which the following conditions exist.

• The capacitors have an initial charge (i.e., an initial voltage) not equal to zero.

• The inductors have an initial magnetic field (i.e., an initial current) not equal to zero.

• Dependent sources are used.

24. KIRCHHOFF'S VOLTAGE LAW

step 1: Identify the loop.

step 2: Pick a loop direction.

step 3: Assign the loop current in the direction picked in step 2.

step 4: Assign voltage polarities consistent with the loop current direction in step 3.

step 5: Apply KVL to the loop using Ohm's law to express the voltages across each circuit element.

step 6: Solve the equation for the desired quantity.

25. KIRCHHOFF'S CURRENT LAW

step 1: Identify the nodes and pick a reference or datum node.

step 2: Label the node-to-datum voltage for each unknown node.

step 3: Pick a current direction for each path at every node.

step 4: Apply KCL to the nodes using Ohm's law to express the currents through each circuit branch.

step 5: Solve the equations for the desired quantity.

26. LOOP ANALYSIS

step 1: Select $n - 1$ loops, that is, one loop less than the total number of possible loops.

step 2: Assign current directions for the selected loops. Show the direction of the current with an arrow.

step 3: Write Kirchhoff's voltage law for each of the selected loops. Assign polarities based on the direction of the loop current. Where two loop currents flow through an element, they are summed to determine the voltage drop in that element.

step 4: Solve the $n - 1$ equations from step 3 for the unknown currents.

step 5: If required, determine the actual current in an element by summing the loop currents flowing through the element.

27. NODE ANALYSIS

step 1: Simplify the circuit, if possible, by combining resistors in series or parallel or by combining current sources in parallel. Identify all nodes. The minimum number of equations required will be $n-1$ where n represents the number of principal nodes.

step 2: Choose one node as the reference node, that is, the node that will be assumed to have ground potential (0 V). To minimize the number of terms in the equations, select the node with the largest number of circuit elements to serve as the reference node.

step 3: Write Kirchhoff's current law for each principal node except the reference node, which is assumed to have a zero potential.

step 4: Solve the $n-1$ equations from step 3 to determine the unknown voltages.

step 5: If required, use the calculated node voltages to determine any branch current desired.

29. VOLTAGE AND CURRENT DIVIDERS

$$\mathbf{V}_2 = \mathbf{V}_s \left(\frac{\mathbf{Z}_2}{\mathbf{Z}_1 + \mathbf{Z}_2} \right) \qquad 29.38$$

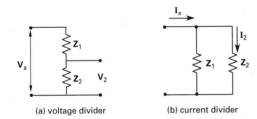

(a) voltage divider   (b) current divider

Figure 29.14 Divider Circuits

$$\mathbf{I}_2 = \mathbf{I}_s \left(\frac{\mathbf{Z}_1}{\mathbf{Z}_1 + \mathbf{Z}_2} \right) = \mathbf{I}_s \left(\frac{\mathbf{G}_2}{\mathbf{G}_1 + \mathbf{G}_2} \right) \qquad 29.40$$

30. STEADY-STATE AND TRANSIENT IMPEDANCE ANALYSIS

DC Steady-State Impedance Analysis

• *resistance*

$$Z_R|_{\mathrm{DC}} = R \qquad 29.41$$

• *inductance*

$$v = L \left(\frac{di}{dt} \right) = L(0) = 0 \qquad 29.42$$

$$Z = \frac{v}{i} = \frac{0}{i} = 0 \qquad 29.43$$

$$Z_L|_{\mathrm{DC}} = 0 \quad [\text{short circuit}] \qquad 29.44$$

• *capacitance*

$$i = C \left(\frac{dv}{dt} \right) = C(0) = 0 \qquad 29.45$$

$$Z = \frac{v}{i} = \frac{v}{0} \to \infty \qquad 29.46$$

$$Z_C|_{\mathrm{DC}} = \infty \quad [\text{open circuit}] \qquad 29.47$$

AC Steady-State Impedance Analysis

• *resistance*

$$Z_R|_{\mathrm{AC}} = R \qquad 29.48$$

• *inductance*

$$v = L \left(\frac{di}{dt} \right) = j\omega L i \qquad 29.49$$

$$Z = \left(\frac{v}{i} \right) = \frac{j\omega L i}{i} = j\omega L \qquad 29.50$$

$$Z_L|_{\mathrm{AC}} = j\omega L \qquad 29.51$$

• *capacitance*

$$i = C \left(\frac{dv}{dt} \right) = C j\omega v \qquad 29.52$$

$$Z = \frac{v}{i} = \frac{v}{C j\omega v} = \frac{1}{j\omega C} \qquad 29.53$$

$$Z_C|_{\mathrm{AC}} = \frac{1}{j\omega C} \qquad 29.54$$

Transient Impedance Analysis

Transient impedance analysis is based on the phasor form with the complex variable s substituted for $j\omega$. The variable $s = \sigma + j\omega$ and is the same as the Laplace transform variable.

• *resistance*

$$Z_R = R \qquad 29.55$$

• *inductance*

$$Z_L = sL \qquad 29.56$$

• *capacitance*

$$Z_C = \frac{1}{sC} \qquad 29.57$$

31. TWO-PORT NETWORKS

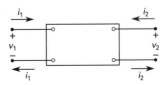

Figure 29.15 Two-Port Network

Table 29.3 *Two-Port Network Parameters*

representation	deriving equations			
hybrid $\begin{bmatrix} h_{11} & h_{12} \\ h_{21} & h_{22} \end{bmatrix}$	$h_{11} = \dfrac{V_1}{I_1}$ $V_2 = 0$	$h_{12} = \dfrac{V_1}{V_2}$ $I_1 = 0$	$h_{21} = \dfrac{I_2}{I_1}$ $V_2 = 0$	$h_{22} = \dfrac{I_2}{V_2}$ $I_1 = 0$
inverse hybrid $\begin{bmatrix} g_{11} & g_{12} \\ g_{21} & g_{22} \end{bmatrix}$	$g_{11} = \dfrac{I_1}{V_1}$ $I_2 = 0$	$g_{12} = \dfrac{I_1}{I_2}$ $V_1 = 0$	$g_{21} = \dfrac{V_2}{V_1}$ $I_2 = 0$	$g_{22} = \dfrac{V_2}{I_2}$ $V_1 = 0$
transmission or chain $\begin{bmatrix} A & B \\ C & D \end{bmatrix}$	$A = \dfrac{V_1}{V_2}$ $I_2 = 0$	$B = \dfrac{-V_1}{I_2}$ $V_2 = 0$	$C = \dfrac{I_1}{V_2}$ $I_2 = 0$	$D = \dfrac{-I_1}{I_2}$ $V_2 = 0$

EERM Chapter 30
Transient Analysis

Chapter, section, equation, figure, and table numbers correspond to EERM. For additional study material, go to the corresponding chapter and section number in EERM.

1. FUNDAMENTALS

$$f(t) = b\left(\frac{dx}{dt}\right) + cx \qquad \textbf{30.1}$$

The solution to Eq. 30.1 is of the general form

$$x(t) = \kappa + Ae^{-\frac{t}{\tau}} \qquad \textbf{30.2}$$

- *time constant,* τ

$$\tau = \frac{b}{c} \qquad \textbf{30.3}$$

- *time constant for an RC circuit*

$$\tau = RC \qquad \textbf{30.4}$$

- *time constant for an RL circuit*

$$\tau = \frac{L}{R} \qquad \textbf{30.5}$$

Table 30.1 *Transient Response*

type of circuit	response
series RC, charging	
$\tau = RC$ $e^{-N} = e^{-\frac{t}{\tau}} = e^{-\frac{t}{RC}}$	$V_{\text{bat}} = v_R(t) + v_C(t)$ $i(t) = \left(\dfrac{V_{\text{bat}} - V_0}{R}\right)e^{-N}$ $v_R(t) = i(t)R$ $\quad = (V_{\text{bat}} - V_0)e^{-N}$ $v_C(t) = V_0 + (V_{\text{bat}} - V_0)$ $\quad \times (1 - e^{-N})$ $Q_C(t) = C\Big(V_0 + (V_{\text{bat}} - V_0)$ $\quad \times (1 - e^{-N})\Big)$
series RC, discharging	
$\tau = RC$ $e^{-N} = e^{-\frac{t}{\tau}} = e^{-\frac{t}{RC}}$	$0 = v_R(t) + v_C(t)$ $i(t) = \left(\dfrac{V_0}{R}\right)e^{-N}$ $v_R(t) = -V_0 e^{-N}$ $v_C(t) = V_0 e^{-N}$ $Q_C(t) = C V_0 e^{-N}$
series RL, charging	
$\tau = \dfrac{L}{R}$ $e^{-N} = e^{-\frac{t}{\tau}} = e^{-\frac{t}{L/R}}$	$V_{\text{bat}} = v_R(t) + v_L(t)$ $i(t) = I_0 e^{-N}$ $\quad + \left(\dfrac{V_{\text{bat}}}{R}\right)\left(1 - e^{-N}\right)$ $v_R(t) = i(t)R$ $\quad = I_0 R e^{-N}$ $\quad + V_{\text{bat}}(1 - e^{-N})$ $v_L(t) = (V_{\text{bat}} - I_0 R)e^{-N}$
series RL, discharging	
$\tau = \dfrac{L}{R}$ $e^{-N} = e^{-\frac{t}{\tau}} = e^{-\frac{t}{L/R}}$	$0 = v_R(t) + v_L(t)$ $i(t) = I_0 e^{-N}$ $v_R(t) = I_0 R e^{-N}$ $v_L(t) = -I_0 R e^{-N}$

10. RESONANT CIRCUITS

A *resonant circuit* has a zero current phase angle difference. The frequency at which the circuit becomes purely resistive is the *resonant frequency*.

The *quality factor*, Q, for a circuit is a dimensionless ratio that compares the reactive energy stored in an inductor each cycle to the resistive energy dissipated.

$$Q = 2\pi \left(\frac{\text{maximum energy stored per cycle}}{\text{energy dissipated per cycle}} \right)$$

$$= \frac{f_0}{(\text{BW})_{\text{Hz}}} = \frac{\omega_0}{(\text{BW})_{\text{rad/s}}}$$

$$= \frac{f_0}{f_2 - f_1} = \frac{\omega_0}{\omega_2 - \omega_1} \quad \begin{bmatrix} \text{parallel} \\ \text{or series} \end{bmatrix} \qquad 30.25$$

- *energy stored in the inductor of a series RLC circuit each cycle*

$$U = \frac{I_m^2 L}{2} = I^2 L = Q \left(\frac{I^2 R}{2\pi f_0} \right) \qquad 30.26$$

- *relationships between the half-power points and quality factor*

$$f_1, f_2 = f_0 \left(\sqrt{1 + \frac{1}{4Q^2}} \mp \frac{1}{2Q} \right)$$

$$\approx f_0 \mp \frac{f_0}{2Q} = f_0 \mp \frac{\text{BW}}{2} \qquad 30.27$$

11. SERIES RESONANCE

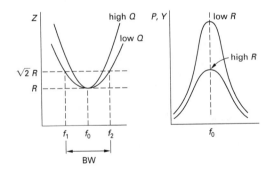

Figure 30.9 *Series Resonance (Band-Pass Filter)*

At the resonant frequency,

$$X_L = X_C \quad \text{[at resonance]} \qquad 30.28$$

$$\omega_0 L = \frac{1}{\omega_0 C} \qquad 30.29$$

$$\omega_0 = 2\pi f_0 = \frac{1}{\sqrt{LC}} \qquad 30.30$$

- *quality factor for a series RLC circuit*

$$Q = \frac{X}{R} = \frac{\omega_0 L}{R}$$

$$= \frac{1}{\omega_0 RC} = \frac{1}{R} \sqrt{\frac{L}{C}}$$

$$= \frac{\omega_0}{(\text{BW})_{\text{rad/s}}} = \frac{f_0}{(\text{BW})_{\text{Hz}}}$$

$$= G\omega_0 L = \frac{G}{\omega_0 C} \qquad 30.32$$

12. PARALLEL RESONANCE

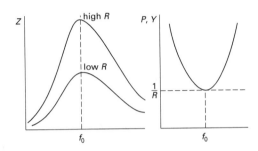

Figure 30.10 *Parallel Resonance (Band-Reject Filter)*

At resonance,

$$X_L = X_C \qquad 30.33$$

$$\omega_0 L = \frac{1}{\omega_0 C} \qquad 30.34$$

$$\omega_0 = 2\pi f_0 = \frac{1}{\sqrt{LC}} \qquad 30.35$$

- *quality factor for a parallel RLC circuit*

$$Q = \frac{R}{X} = \omega_0 RC = \frac{R}{\omega_0 L}$$

$$= R \sqrt{\frac{C}{L}} = \frac{\omega_0}{(\text{BW})_{\text{rad/s}}} = \frac{f_0}{(\text{BW})_{\text{Hz}}}$$

$$= \frac{\omega_0 C}{G} = \frac{1}{G\omega_0 L} \qquad 30.37$$

EERM Chapter 31
Time Response

> Chapter, section, equation, figure, and table numbers correspond to EERM. For additional study material, go to the corresponding chapter and section number in EERM.

2. FIRST-ORDER ANALYSIS

$$f(t) = b\frac{dx}{dt} + cx \qquad 31.2$$

$$x(t) = \kappa + Ae^{-\frac{t}{\tau}} \qquad 31.3$$

A capacitive first-order circuit generally uses voltage as the dependent variable.

$$v_{\text{Th}} = R_{\text{Th}}C\left(\frac{dv_C}{dt}\right) + v_C \qquad 31.5$$

The solution is

$$v_C(t) = V_{C,\text{ss}} + Ae^{-\frac{t}{\tau}} \qquad 31.6$$

An inductive first-order circuit generally uses current as the dependent variable.

$$v_{\text{Th}} = L\left(\frac{di_L}{dt}\right) + R_{\text{Th}}i_L \qquad 31.8$$

The solution is

$$i_L(t) = I_{L,\text{ss}} + Ae^{-\frac{t}{\tau}} \qquad 31.9$$

5. SECOND-ORDER ANALYSIS

The differential equation for second-order circuits takes on the general form

$$f(t) = a\frac{d^2x}{dt^2} + b\frac{dx}{dt} + cx \qquad 31.13$$

The associated characteristic equation, written with the roots shown in the s domain, is

$$as^2 + bs + c = 0 \qquad 31.14$$

6. SECOND-ORDER ANALYSIS: OVERDAMPED

If the two roots of Eq. 31.14 are real and different from one another, equivalent to the condition $b^2 > 4ac$, the solution is of the form

$$x(t) = \kappa + Ae^{s_1 t} + Be^{s_2 t} \qquad 31.15$$

$$s_1 = \frac{-b + \sqrt{b^2 - 4ac}}{2a} \qquad 31.16$$

$$s_2 = \frac{-b - \sqrt{b^2 - 4ac}}{2a} \qquad 31.17$$

$$A = \left(\frac{1}{2}\right)\left(1 + \frac{b}{\sqrt{b^2 - 4ac}}\right)\left(x(0^+) - x_{\text{ss}}\right)$$

$$+ \left(\frac{a}{\sqrt{b^2 - 4ac}}\right)\left(x'(0^+) - x'_{\text{ss}}\right) \qquad 31.24$$

$$B = \left(\frac{1}{2}\right)\left(1 - \frac{b}{\sqrt{b^2 - 4ac}}\right)\left(x(0^+) - x_{\text{ss}}\right)$$

$$- \left(\frac{a}{\sqrt{b^2 - 4ac}}\right)\left(x'(0^+) - x'_{\text{ss}}\right) \qquad 31.25$$

7. SECOND-ORDER ANALYSIS: CRITICALLY DAMPED

If the two roots of Eq. 31.14 are real and the same, equivalent to the condition $b^2 = 4ac$, the solution is of the form

$$x(t) = \kappa + Ae^{st} + Bte^{st} \qquad 31.26$$

$$s = \frac{-b}{2a} \qquad 31.27$$

$$A = x(0^+) - x_{\text{ss}} \qquad 31.34$$

$$B = x'(0^+) - x'_{\text{ss}} + \left(\frac{b}{2a}\right)A \qquad 31.35$$

8. SECOND-ORDER ANALYSIS: UNDERDAMPED

If the two roots of Eq. 31.14 are complex conjugates, equivalent to the condition $b^2 < 4ac$, the solution is of the form

$$x(t) = \kappa + e^{\alpha t}\left(A_e e^{j\beta t} + B_e e^{-j\beta t}\right) \qquad 31.36$$

This can also be represented in the sinusoidal form as

$$x(t) = \kappa + e^{\alpha t}\left(A\cos\beta t + B\sin\beta t\right) \qquad 31.37$$

$$s_1 = \alpha + j\beta \qquad 31.38$$

$$s_2 = \alpha - j\beta \qquad 31.39$$

The roots of the characteristic equation are

$$\alpha = \frac{-b}{2a} \qquad 31.40$$

$$\beta = \frac{\sqrt{4ac - b^2}}{2a} \qquad 31.41$$

$$A = x(0^+) - x_{\text{ss}} \qquad 31.48$$

$$B = \left(\frac{\alpha}{\beta}\right)A + \left(\frac{1}{\beta}\right)\left(x'(0^+) - x'_{\text{ss}}\right) \qquad 31.49$$

11. COMPLEX FREQUENCY

- *complex frequency*

$$\mathbf{s} = \sigma + j\omega \qquad 31.74$$

Table 31.4 *Generalized Impedance in the s Domain*

impedance type	frequency domain value	s domain value
Z_R	R	R
Z_L	$j\omega L$	sL
Z_C	$\dfrac{1}{j\omega C}$	$\dfrac{1}{sC}$

12. LAPLACE TRANSFORM ANALYSIS

- *Laplace transform*

$$\mathcal{L}\{f(t)\} = \int\limits_{0^+}^{\infty} f(t)e^{-st}\,dt \qquad 31.77$$

$$\mathcal{L}\{f(t)\} = F(s) \qquad 31.78$$

$$\mathcal{L}\left\{\frac{d^2 f(t)}{dt^2}\right\} = s^2 F(s) - sf(0^+) - f'(0^+) \qquad 31.79$$

$$\mathcal{L}\left\{\frac{df(t)}{dt}\right\} = sF(s) - f(0^+) \qquad 31.80$$

- *inverse Laplace transform*

$$\mathcal{L}^{-1}\{F(s)\} = f(t) = \frac{1}{2\pi j}\int\limits_{\sigma-j\infty}^{\sigma+j\infty} F(s)e^{st}\,dt \qquad 31.81$$

The classical method is useful when the desired quantity is one of the state variables, capacitor voltage or inductor current. If the desired variables are other than the state variables, the *circuit transformation method* is the most direct.

13. CAPACITANCE IN THE *s* DOMAIN

$$\mathcal{L}\{i_C(t)\} = \mathcal{L}\left\{C\frac{dv_C(t)}{dt}\right\}$$

$$= sCV_C(s) - Cv_C(0) \qquad 31.83$$

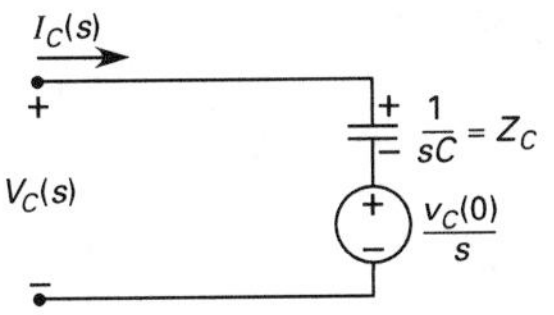

Figure 31.9 Capacitor Impedance and Initial Voltage Source Model

14. INDUCTANCE IN THE *s* DOMAIN

$$\mathcal{L}\{v_L(t)\} = \mathcal{L}\left\{L\frac{di_L(t)}{dt}\right\}$$

$$= sLI_L(s) - Li_L(0) \qquad 31.85$$

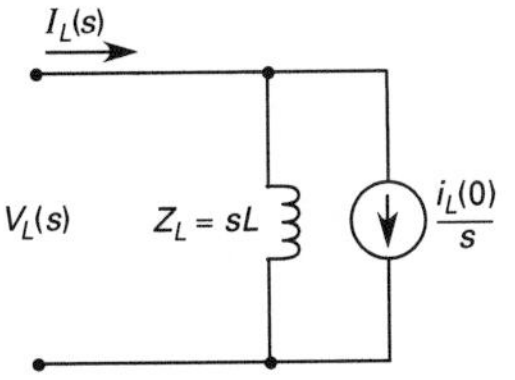

Figure 31.12 Inductor Impedance and Initial Current Source Model

EERM Chapter 32
Frequency Response

> Chapter, section, equation, figure, and table numbers correspond to EERM. For additional study material, go to the corresponding chapter and section number in EERM.

2. TRANSFER FUNCTION

The term "transfer function" refers to the relationship of one electrical parameter in a network to a second electrical parameter elsewhere in the network.

$$\mathbf{T}_{\text{net}}(\mathbf{s}) = \frac{\mathbf{Y}(\mathbf{s})}{\mathbf{X}(\mathbf{s})}$$

$$= A\left(\frac{(\mathbf{s}-\mathbf{z}_1)(\mathbf{s}-\mathbf{z}_2)\dots(\mathbf{s}-\mathbf{z}_n)}{(\mathbf{s}-\mathbf{p}_1)(\mathbf{s}-\mathbf{p}_2)\dots(\mathbf{s}-\mathbf{p}_d)}\right) \qquad 32.2$$

The complex constants represented as $\mathbf{z}_y$ $(y = 1, 2, \dots, n)$ are the *zeros* of $\mathbf{T}_{\text{net}}(\mathbf{s})$ and are plotted in the **s** domain as $\bigcirc$'s. The complex constants represented as $\mathbf{p}_x$ $(x = 1, 2, \dots, d)$ are the *poles* of $\mathbf{T}_{\text{net}}(\mathbf{s})$ and are plotted in the **s** domain as X's.

3. STEADY-STATE RESPONSE

$$(\mathbf{s}-\mathbf{z}_n) = N_n\angle\alpha_n \qquad 32.3$$

$$(\mathbf{s}-\mathbf{p}_d) = D_d\angle\beta_d \qquad 32.4$$

$$\mathbf{T}_{\text{net}}(\mathbf{s}) = A\frac{(N_1\angle\alpha_1)(N_2\angle\alpha_2)\cdots(N_n\angle\alpha_n)}{(D_1\angle\beta_1)(D_2\angle\beta_2)\cdots(D_d\angle\beta_d)}$$

$$= A\left(\frac{N_1 N_2 \cdots N_n}{D_1 D_2 \dots D_d}\right)\angle(\alpha_1 + \alpha_2 + \cdots + \alpha_n)$$

$$- (\beta_1 + \beta_2 + \cdots \beta_d) \qquad 32.5$$

4. TRANSIENT RESPONSE

The transient response is characterized by the poles of the network function.

6. BODE PLOT PRINCIPLES: MAGNITUDE PLOT

A *Bode plot* or *Bode diagram* is a plot of the gain or phase of an electrical device or network against the frequency.

- *gain* (any transfer function's magnitude)

$$G = 20\log|T(s)| \qquad 32.19$$

- *transfer function for a single zero at the origin*

$$T(s) = \sigma_0 \left(\frac{s}{\sigma_0} \right) \qquad 32.20$$

$$G = 20 \log \left| \sigma_0 \left(\frac{j\omega}{\sigma_0} \right) \right|$$

$$= 20 \log \sigma_0 + 20 \log \left(\frac{\omega}{\sigma_0} \right) \qquad 32.21$$

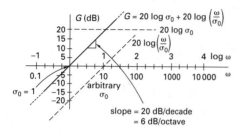

Figure 32.5 *Bode Plot: Single Zero at Origin*

- *transfer function for a single pole at the origin*

$$T(s) = \frac{1}{\sigma_0 \left(\dfrac{s}{\sigma_0} \right)} \qquad 32.24$$

- *gain with $s = j\omega$*

$$G = 20 \log \left| \frac{1}{\sigma_0 \left(\dfrac{j\omega}{\sigma_0} \right)} \right|$$

$$= -20 \log \sigma_0 - 20 \log \left(\frac{\omega}{\sigma_0} \right) \qquad 32.25$$

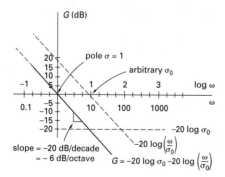

Figure 32.6 *Bode Plot: Single Pole at Origin*

- *transfer function for a single zero on the negative real axis at $-\sigma_0$ in the s plane*

$$T(s) = \sigma_0 \left(1 + \frac{s}{\sigma_0} \right) \qquad 32.26$$

- *gain with $s = j\omega$*

$$G = 20 \log \left| \sigma_0 \left(1 + \frac{j\omega}{\sigma_0} \right) \right|$$

$$= 20 \log \sigma_0 + 20 \log \left| \left(1 + \frac{\omega}{\sigma_0} \right) \right| \qquad 32.27$$

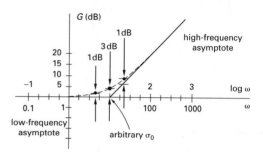

Figure 32.7 *Bode Plot: Single Zero on Negative Real Axis*

- *transfer function for a single pole on the negative real axis at $-\sigma_0$ in the s plane*

$$T = \sigma_0 \left(\frac{1}{1 + \dfrac{s}{\sigma_0}} \right)$$

- *gain with $s = j\omega$*

$$G = 20 \log \left| \left(\frac{1}{\sigma_0 \left(1 + \dfrac{j\omega}{\sigma_0} \right)} \right) \right|$$

$$= -20 \log \sigma_0 - 20 \log \left| \left(1 + \frac{\omega}{\sigma_0} \right) \right| \qquad 32.32$$

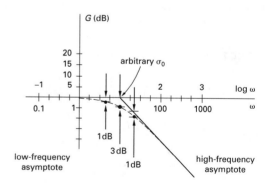

Figure 32.8 *Bode Plot: Single Pole on Negative Real Axis*

7. BODE PLOT PRINCIPLES: PHASE PLOT

$$\theta_z = \tan^{-1}\left(\frac{\omega}{\sigma_{zn}}\right) \quad \text{[radians]} \qquad \textit{32.35}$$

$$\theta_p = -\tan^{-1}\left(\frac{\omega}{\sigma_{pd}}\right) \quad \text{[radians]} \qquad \textit{32.36}$$

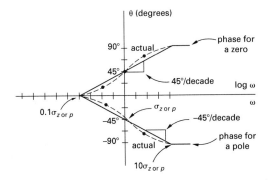

Figure 32.9 *Bode Phase Plot for a Zero or Pole*

8. BODE PLOT METHODS

- *general form of the transfer function of a circuit*

$$T(s) = A\left(\frac{\begin{array}{c}(s+z_1)(s+z_2)\\ \cdots \times (s+z_n)(s)^y\end{array}}{\begin{array}{c}(s+p_1)(s+p_2)\\ \cdots \times (s+p_d)(s)^x\end{array}}\right) \qquad \textit{32.39}$$

Using the normalized notation changes Eq. 32.39 to Eq. 32.40.

$$T(j\omega) = A\left(\frac{z_1 z_2 \cdots z_n}{p_1 p_2 \cdots p_d}\right)$$

$$\times \left(\frac{\begin{array}{c}\left(1+\dfrac{j\omega}{z_1}\right)\left(1+\dfrac{j\omega}{z_2}\right)\\ \cdots \times \left(1+\dfrac{j\omega}{z_n}\right)(j\omega)^y\end{array}}{\begin{array}{c}\left(1+\dfrac{j\omega}{p_1}\right)\left(1+\dfrac{j\omega}{p_2}\right)\\ \cdots \times \left(1+\dfrac{j\omega}{p_d}\right)(j\omega)^x\end{array}}\right) \qquad \textit{32.40}$$

Combining the constant term A with the values of the zeros and poles gives a new constant, K.

$$T(j\omega) = K\left(\frac{\begin{array}{c}\left(1+\dfrac{j\omega}{z_1}\right)\left(1+\dfrac{j\omega}{z_2}\right)\\ \cdots \times \left(1+\dfrac{j\omega}{z_n}\right)(j\omega)^y\end{array}}{\begin{array}{c}\left(1+\dfrac{j\omega}{p_1}\right)\left(1+\dfrac{j\omega}{p_2}\right)\\ \cdots \times \left(1+\dfrac{j\omega}{p_d}\right)(j\omega)^x\end{array}}\right) \qquad \textit{32.41}$$

The constant K is

$$K = A\frac{\displaystyle\prod_1^n z_n}{\displaystyle\prod_1^d p_d} \qquad \textit{32.42}$$

$$G = 20\log K + \sum_{n=1}^{n} 20\log\left|\left(1+\frac{j\omega}{z_n}\right)\right| + 20y\log\omega$$

$$-\sum_{d=1}^{d} 20\log\left|\left(1+\frac{j\omega}{p_d}\right)\right| - 20x\log\omega \qquad \textit{32.43}$$

The angle of the gain, θ_G, in degrees is

$$\theta_G = \sum_{n=1}^{n} \tan^{-1}\left(\frac{\omega}{z_n}\right) + 90y$$

$$-\sum_{d=1}^{d} \tan^{-1}\left(\frac{\omega}{p_d}\right) - 90x \qquad \textit{32.44}$$

Power—Generation

EERM Chapter 33
Generation Systems

> Chapter, section, equation, figure, and table numbers correspond to EERM. For additional study material, go to the corresponding chapter and section number in EERM.

6. ALTERNATING CURRENT GENERATORS

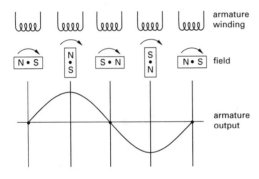

Figure 33.8 *AC Output Generation*

$$N_s = \frac{120f}{p} \qquad 33.5$$

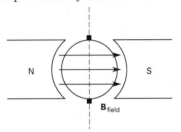

Figure 33.9 *Real Load Sharing*

$$P = \frac{f_{nl} - f_{sys}}{f_{droop}} \qquad 33.8$$

$$Q = \frac{V_{sys} - V_{nl}}{V_{droop}} \qquad 33.9$$

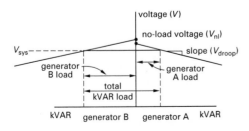

Figure 33.10 *Reactive Load Sharing*

If the droop is given or used as a positive value, Eq. 33.9 becomes

$$Q = \frac{V_{nl} - V_{sys}}{V_{droop}} \qquad 33.10$$

8. DIRECT CURRENT GENERATORS

Commutation is the process of current reversal in the armature windings that provides direct current to the brushes.

Armature reaction is the interaction between the magnetic flux produced by the armature current and the magnetic flux produced by the field current.

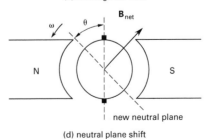

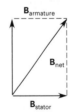

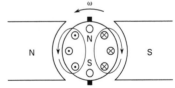

Figure 33.13 *DC Generator Armature Flux*

EERM Chapter 34
Three-Phase Electricity and Power

> Chapter, section, equation, figure, and table numbers correspond to EERM. For additional study material, go to the corresponding chapter and section number in EERM.

2. GENERATION OF THREE-PHASE POTENTIAL

$$\mathbf{V}_a = V_p \angle 0° \qquad 34.1$$

$$\mathbf{V}_b = V_p \angle -120° \qquad 34.2$$

$$\mathbf{V}_c = V_p \angle -240° \qquad 34.3$$

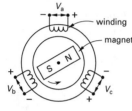

(a) alternator

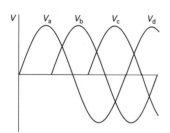

(b) ABC (positive) sequence

Figure 34.1 *Three-Phase Voltage*

5. DELTA-CONNECTED LOADS

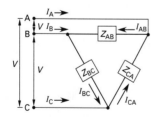

Figure 34.4 *Delta-Connected Loads*

$$|\mathbf{I}_A| = |\mathbf{I}_{AB} - \mathbf{I}_{CA}| = \sqrt{3}I_{AB} \qquad 34.13$$

$$|\mathbf{I}_B| = |\mathbf{I}_{BC} - \mathbf{I}_{AB}| = \sqrt{3}I_{BC} \qquad 34.14$$

$$|\mathbf{I}_C| = |\mathbf{I}_{CA} - \mathbf{I}_{BC}| = \sqrt{3}I_{CA} \qquad 34.15$$

$$P_t = 3P_p = 3V_p I_p \cos\phi = \sqrt{3}VI \cos\phi \qquad 34.17$$

6. WYE-CONNECTED LOADS

$$\mathbf{I}_A = \mathbf{I}_{AN} = \frac{\mathbf{V}_{AN}}{\mathbf{Z}_{AN}} = \frac{\mathbf{V}}{\sqrt{3}\mathbf{Z}_{AN}} \qquad 34.18$$

$$\mathbf{I}_B = \mathbf{I}_{BN} = \frac{\mathbf{V}_{BN}}{\mathbf{Z}_{BN}} = \frac{\mathbf{V}}{\sqrt{3}\mathbf{Z}_{BN}} \qquad 34.19$$

$$\mathbf{I}_C = \mathbf{I}_{CN} = \frac{\mathbf{V}_{CN}}{\mathbf{Z}_{CN}} = \frac{\mathbf{V}}{\sqrt{3}\mathbf{Z}_{CN}} \qquad 34.20$$

$$\mathbf{I}_N = 0 \quad \text{[balanced]} \qquad 34.21$$

$$P_t = 3P_p = 3V_p I_p \cos\phi = \sqrt{3}VI \cos\phi \qquad 34.22$$

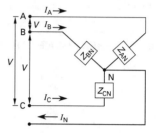

Figure 34.5 *Wye-Connected Loads*

8. PER-UNIT CALCULATIONS

$$\text{per unit} = \frac{\text{actual}}{\text{base}} = \frac{\text{percent}}{100\%} \qquad 34.23$$

For a three-phase system, the usual bases are the line voltage (in kV) and the total (three-phase) apparent power (in kVA) ratings.

$$V_p = \frac{V_l}{\sqrt{3}} \qquad 34.24$$

$$S_p = VA_p = \frac{S_t}{3} = \frac{VA_t}{3} \qquad 34.25$$

The per-unit system is represented by Eqs. 34.26 through 34.31 and is presented elsewhere as phase bases for use with single-phase systems. Conversion to a three-phase base is accomplished with Eqs. 34.24 and 34.25.

$$S_{\text{base}} = S_p \qquad 34.26$$

$$V_{\text{base}} = V_p \qquad 34.27$$

$$I_{\text{base}} = \frac{S_{\text{base}}}{V_{\text{base}}} = \frac{S_p}{V_p} \qquad 34.28$$

$$Z_{\text{base}} = \frac{V_{\text{base}}}{I_{\text{base}}} = \frac{V_p^2}{S_p} \qquad 34.29$$

$$P_{\text{base}} = S_p \qquad 34.30$$

$$Q_{\text{base}} = S_p \qquad 34.31$$

$$I_{\text{pu}} = \frac{I_{\text{actual}}}{I_{\text{base}}} \qquad 34.32$$

$$V_{\text{pu}} = \frac{V_{\text{actual}}}{V_{\text{base}}} \qquad 34.33$$

$$Z_{\text{pu}} = \frac{Z_{\text{actual}}}{Z_{\text{base}}} \qquad 34.34$$

$$P_{\text{pu}} = \frac{P_{\text{actual}}}{P_{\text{base}}} \qquad 34.35$$

$$Q_{\text{pu}} = \frac{Q_{\text{actual}}}{Q_{\text{base}}} \qquad 34.36$$

Ohm's law and other circuit analysis methods can be used with the per-unit quantities.

$$V_{\text{pu}} = I_{\text{pu}} Z_{\text{pu}} \qquad 34.37$$

The general method for converting from one per-unit value, call it χ, to another in a different base is

$$\chi_{\text{pu,new}} = \chi_{\text{pu,old}} \left(\frac{\chi_{\text{base,old}}}{\chi_{\text{base,new}}} \right) \qquad 34.38$$

The impedance per-unit value conversion is

$$Z_{\text{pu,new}} = Z_{\text{pu,old}} \left(\frac{V_{\text{base,old}}}{V_{\text{base,new}}} \right)^2 \left(\frac{S_{\text{base,new}}}{S_{\text{base,old}}} \right) \qquad 34.39$$

9. UNBALANCED LOADS

Unbalanced systems can be evaluated by computing the phase currents and then applying Kirchhoff's current law (in vector form) to obtain the line currents. The *neutral current* will be

$$\mathbf{I}_N = -(\mathbf{I}_A + \mathbf{I}_B + \mathbf{I}_C) \qquad 34.40$$

Power—Transmission

EERM Chapter 35
Power Distribution

> Chapter, section, equation, figure, and table numbers correspond to EERM. For additional study material, go to the corresponding chapter and section number in EERM.

6. UNDERGROUND DISTRIBUTION

- *optimal thickness*

$$\frac{r_2}{r_1} = e = 2.718 \qquad \textit{35.3}$$

- *operating voltage for a capacitance-graded cable*

$$V = E_{\max}\left(r_1 \ln\left(\frac{r_2}{r_1}\right) + r_2 \ln\left(\frac{r_3}{r_2}\right)\right) \qquad \textit{35.4}$$

7. FAULT ANALYSIS: SYMMETRICAL

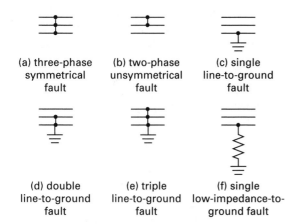

(a) three-phase symmetrical fault

(b) two-phase unsymmetrical fault

(c) single line-to-ground fault

(d) double line-to-ground fault

(e) triple line-to-ground fault

(f) single low-impedance-to-ground fault

Figure 35.7 *Fault Types*

A three-phase *symmetrical fault* has three specific time periods of concern as shown in Fig. 35.8. The voltage E_g'' is calculated for the subtransient interval *just prior to the initiation of the fault* using

$$E_g'' = V_t + jI_L X_d'' \qquad \textit{35.11}$$

During the *transient period*, the correct generator voltage for this period is calculated *just prior to the initiation of the fault* using

$$E_g' = V_t + jI_L X_d' \qquad \textit{35.12}$$

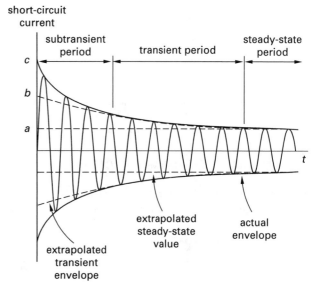

(a) synchronous generator three-phase fault response

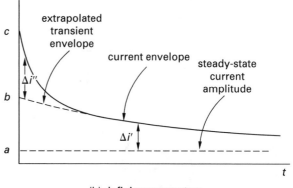

(b) defining parameters

Figure 35.8 *Symmetrical Fault Terminology*

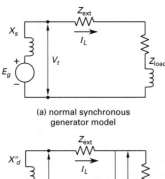

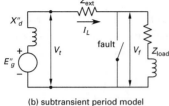

(a) normal synchronous
generator model

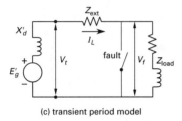

(b) subtransient period model

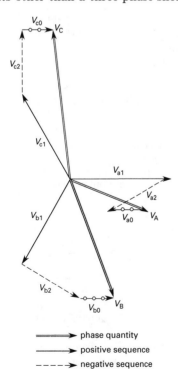

(c) transient period model

Figure 35.9 *Synchronous Generator Fault Models*

8. FAULT ANALYSIS: UNSYMMETRICAL

Unsymmetrical faults, also called *asymmetrical faults*, are any faults other than a three-phase short.

phase quantity
positive sequence
negative sequence
zero sequence

Figure 35.10 *Phasor Diagram: Symmetrical Components of Unbalanced Phasors*

The unsymmetrical phasors are represented in terms of their symmetrical components by the following equations.

$$V_A = V_{a0} + V_{a1} + V_{a2} \qquad 35.14$$

$$V_B = V_{b0} + V_{b1} + V_{b2} \qquad 35.15$$

$$V_C = V_{c0} + V_{c1} + V_{c2} \qquad 35.16$$

Consider an operator **a** defined as a unit vector with a magnitude of one and an angle of $120°$, $1\angle 120°$.

$$
\begin{aligned}
\text{a} &= 1\angle 120° = 1 \times e^{j120°} \\
&= -0.5 + j0.866 \qquad 35.17 \\
\text{a}^2 &= 1\angle 240° = -0.5 - j0.866 = \text{a}^* \quad 35.18 \\
\text{a}^3 &= 1\angle 360° = 1\angle 0° \qquad 35.19 \\
\text{a}^4 &= \text{a} \qquad 35.20 \\
\text{a}^5 &= \text{a}^2 \qquad 35.21 \\
\text{a}^6 &= \text{a}^3 \qquad 35.22 \\
1 + \text{a} + \text{a}^2 &= 0 \qquad 35.23
\end{aligned}
$$

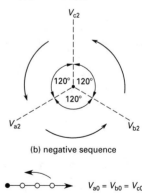

(a) positive sequence

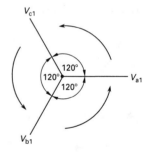

(b) negative sequence

$$V_{a0} = V_{b0} = V_{c0}$$

(c) zero sequence

Figure 35.11 *Components of Unsymmetrical Phasors*

The unsymmetrical components can be represented in terms of a single phase.

$$V_A = V_{a0} + V_{a1} + V_{a2} \qquad 35.24$$

$$V_B = V_{a0} + \text{a}^2 V_{a1} + \text{a}V_{a2} \qquad 35.25$$

$$V_C = V_{a0} + \text{a}V_{a1} + \text{a}^2 V_{a2} \qquad 35.26$$

Solving for the sequence components,

$$V_{a0} = \frac{1}{3}\left(V_a + V_b + V_c\right) \qquad 35.27$$

$$V_{a1} = \frac{1}{3}\left(V_a + aV_b + a^2V_c\right) \qquad 35.28$$

$$V_{a2} = \frac{1}{3}\left(V_a + a^2V_b + aV_c\right) \qquad 35.29$$

EERM Chapter 36
Power Transformers

> Chapter, section, equation, figure, and table numbers correspond to EERM. For additional study material, go to the corresponding chapter and section number in EERM.

1. THEORY

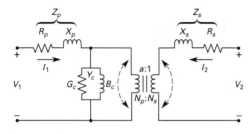

Figure 36.1 *Exact Transformer Model*

2. TRANSFORMER RATING

- *eddy current losses, P_e*
- *hysteresis losses, P_h*
- *mass of iron, m*
- *maximum flux density, B_m*

The exponent n is called the *Steinmetz exponent*.

- *coupling coefficient, k*

$$P_e = k_e B_m^2 f^2 m \qquad 36.1$$

$$P_h = k_h B_m^n fm \qquad 36.2$$

- *core losses*

$$P_c = \frac{V_1^2}{R_c} \qquad 36.3$$

- *copper losses, P_{Cu}*

$$P_{Cu} = I^2 R = I_1^2 R_p + I_2^2 R_s \qquad 36.4$$

- *transformer efficiency*

$$\eta = \frac{P_{out}}{P_{in}} = \frac{P_{in} - \sum P_{losses}}{P_{in}}$$

$$= \frac{P_{out}}{P_{out} + P_c + P_{Cu}} \qquad 36.5$$

3. VOLTAGE REGULATION

$$\text{VR} = \frac{V_{nl} - V_{fl}}{V_{fl}} \qquad 36.6$$

$$\text{VR} = \frac{\dfrac{V_p}{a} - V_{s,\text{rated}}}{V_{s,\text{rated}}} \qquad 36.7$$

4. CONNECTIONS

$$V_l = V_\phi \quad [\text{delta}] \qquad 36.8$$

$$I_l = \sqrt{3}I_\phi \quad [\text{delta}] \qquad 36.9$$

$$V_l = \sqrt{3}V_\phi \quad [\text{wye}] \qquad 36.10$$

$$I_l = I_\phi \quad [\text{wye}] \qquad 36.11$$

$$P = \sqrt{3}I_l V_l \text{ pf} = 3I_\phi V_\phi \text{ pf} \qquad 36.12$$

6. OPEN-CIRCUIT TEST

The open-circuit test determines the core parameters and the turns ratio.

- *admittance*

$$Y_c = G_c + jB_c = \frac{I_{1oc}}{V_{1oc}} \qquad 36.13$$

- *conductance*

$$G_c = \frac{P_{oc}}{V_{1oc}^2} \qquad 36.14$$

- *susceptance*

$$B_c = \frac{1}{X_c} = \frac{-1}{\omega L_c} = -\sqrt{Y_c^2 - G_c^2}$$

$$= \frac{-\sqrt{I_{1oc}^2 V_{1oc}^2 - P_{oc}^2}}{V_{1oc}^2} \qquad 36.15$$

- *turns ratio*

$$a_{ps} = \frac{V_{1oc}}{V_{2oc}} \qquad 36.16$$

- *power*

$$P_{oc} = V_{oc}^2 G_c \qquad 36.17$$

- *reactive power*

$$Q_{oc} = V_{1oc}^2 B_c \qquad 36.18$$

- *apparent power*

$$S_{oc} = V_{1oc}^2 Y_c = V_{1oc} I_{1oc} = \sqrt{P_{oc}^2 + Q_{oc}^2} \qquad 36.19$$

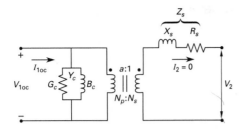

Figure 36.4 *Transformer Open-Circuit Test Model*

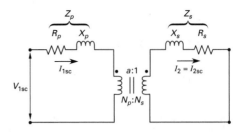

Figure 36.5 *Transformer Short-Circuit Test Model*

7. SHORT-CIRCUIT TEST

The short-circuit test determines the winding impedances and verifies the turns ratio.

$$Z = R_p + jX_p + a_{ps}^2 (R_s + jX_s) = \frac{V_{1sc}}{I_{1sc}} \qquad 36.20$$

- *total resistance, R*

$$R = R_p + a_{ps}^2 R_s = \frac{P_{sc}}{I_{sc}^2} \qquad 36.21$$

To maximize efficiency, transformers are normally designed with R_p equal to $a_{ps}^2 R_s$.

$$R_p = a_{ps}^2 R_s = \frac{P_{sc}}{2I_{sc}^2} \qquad 36.22$$

- *total reactance, X*

$$X = X_p + a_{ps}^2 X_s = \frac{\sqrt{I_{sc}^2 V_{sc}^2 - P_{sc}^2}}{I_{sc}^2} \qquad 36.23$$

To maximize efficiency, transformers are normally designed with X_p equal to $a_{ps}^2 X_s$.

$$X_p = a_{ps}^2 X_s = \frac{Q_{sc}}{2I_{sc}^2} \qquad 36.24$$

- *turns ratio*

$$a_{ps} = \frac{I_{2sc}}{I_{1sc}} \qquad 36.25$$

- *power*

$$P_{sc} = I_{sc}^2 R = I_{sc}^2 \left(R_p + a_{ps}^2 R_s \right) \qquad 36.26$$

- *reactive power*

$$Q_{sc} = I_{sc}^2 X = I_{sc}^2 \left(X_p + a_{ps}^2 X_s \right) \qquad 36.27$$

- *apparent power*

$$S_{sc} = I_{1sc}^2 Z_{sc} = V_{1sc} I_{1sc} = \sqrt{P_{sc}^2 + Q_{sc}^2} \qquad 36.28$$

8. ABCD PARAMETERS

- *ABCD parameters for any two-port network*

$$V_{in} = AV_{out} - BI_{out} \qquad 36.29$$

$$I_{in} = CV_{out} - DI_{out} \qquad 36.30$$

EERM Chapter 37
Power Transmission Lines

Chapter, section, equation, figure, and table numbers correspond to EERM. For additional study material, go to the corresponding chapter and section number in EERM.

1. FUNDAMENTALS

$$Z_0 = \sqrt{\frac{Z_l}{Y_l}} \quad [\text{in } \Omega] \qquad 37.1$$

$$\Gamma = \frac{V_{reflected}}{V_{incident}} = \frac{I_{reflected}}{I_{incident}}$$

$$= \left| \frac{Z_{load} - Z_0}{Z_{load} + Z_0} \right| \qquad 37.3$$

The fraction of incident power that is reflected back to the source from the load is Γ^2.

$$\Gamma = \frac{SWR - 1}{SWR + 1} \qquad 37.4$$

- *velocity of propagation*

$$v_w = \frac{1}{\sqrt{L_l C_l}} \qquad 37.5$$

3. SKIN EFFECT

$$\delta = \frac{1}{\sqrt{\dfrac{\pi f \mu}{\rho}}} = \frac{1}{\sqrt{\pi f \mu \sigma}} \qquad 37.8$$

- *AC resistance per unit length for a flat conducting plate of unit width w*

$$R_{l,\mathrm{AC}} = \frac{\rho}{\delta w} \quad [\text{in } \Omega/\text{m}] \qquad 37.9$$

$$\delta_{\mathrm{Cu}} = \frac{0.066}{\sqrt{f}} \quad [\text{in m}] \qquad 37.10$$

7. SINGLE-PHASE INDUCTANCE

Inductance of a single-phase system, which consists of two conductors, is the sum of the internal and external inductances.

$$L_l = L_{l,\mathrm{int}} + L_{l,\mathrm{ext}}$$

$$= \left(\frac{\mu_0}{4\pi}\right)\left(1 + 4\ln\left(\frac{D}{r}\right)\right) \quad [\text{in H/m}] \qquad 37.19$$

Equation 37.19 is simplified using the concept of the *geometric mean radius* (GMR).

$$\mathrm{GMR} = re^{-\frac{1}{4}} \qquad 37.20$$

$$L_l = \left(\frac{\mu_0}{\pi}\right)\ln\left(\frac{D}{\mathrm{GMR}}\right)$$

$$= 4 \times 10^{-7}\ln\left(\frac{D}{\mathrm{GMR}}\right) \quad [\text{in H/m}] \qquad 37.21$$

8. SINGLE-PHASE CAPACITANCE

$$C_l = \frac{\pi\varepsilon_0}{\ln\left(\frac{D}{r}\right)} = \frac{2.78 \times 10^{-11}}{\ln\left(\frac{D}{r}\right)} \quad [\text{in F/m}] \qquad 37.22$$

9. THREE-PHASE TRANSMISSION

When the conductors are not symmetrically arranged, as is often the case, formulas used for the inductance and capacitance are still valid if the equivalent distance, D_e, is substituted for the distance.

$$D_e = \sqrt[3]{D_{\mathrm{ab}}D_{\mathrm{bc}}D_{\mathrm{ca}}} \qquad 37.23$$

- *per-phase inductance per unit length of a three-phase transmission line*

$$L_l = \frac{\mu_0}{2\pi}\ln\left(\frac{D_e}{\mathrm{GMR}}\right)$$

$$= 2 \times 10^{-7}\ln\left(\frac{D_e}{\mathrm{GMR}}\right) \quad [\text{in H/m}] \qquad 37.24$$

- *per-phase capacitance per unit length of a three-transmission line*

$$C_l = \frac{2\pi\varepsilon_0}{\ln\left(\dfrac{D_e}{\mathrm{GMR}}\right)} = \frac{5.56 \times 10^{-11}}{\ln\left(\dfrac{D_e}{\mathrm{GMR}}\right)} \quad [\text{in F/m}] \quad 37.25$$

10. POWER TRANSMISSION LINES

$$\mathbf{V}_l = \mathbf{I}R_l + j\mathbf{I}X_l = \mathbf{I}Z_l \qquad 37.26$$

$$X_L = 2.022 \times 10^{-3}f\ln\left(\frac{D}{\mathrm{GMR}}\right) \quad [\text{in } \Omega/\text{mi}] \quad 37.27$$

For line-to-line spacings other than one foot, the correction factor given by Eq. 37.28 must be applied.

$$K_L = 1 + \frac{\ln D}{\ln\left(\dfrac{1}{\mathrm{GMR}}\right)} \qquad 37.28$$

- *formula for capacitive reactance per mile (approximately)*

$$X_C = \left(\frac{1.781 \times 10^6}{f}\right)\ln\left(\frac{D}{r}\right) \quad [\text{in } \Omega/\text{mi}] \quad 37.29$$

For line-to-line spacings other than one foot, the correction factor given by Eq. 37.30 must be applied.

$$K_C = 1 + \frac{\ln D}{\ln\left(\dfrac{1}{r}\right)} \qquad 37.30$$

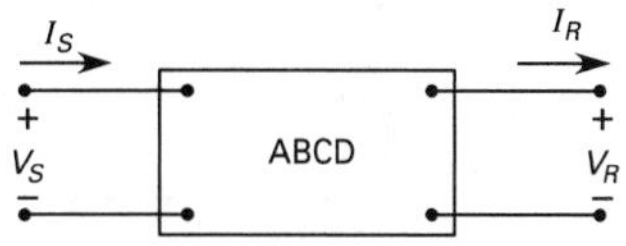

(a) network

$$V_S = AV_R + BI_R$$
$$I_S = CV_R + DI_R$$
(b) equations

$$\begin{bmatrix} V_S \\ I_S \end{bmatrix} = \begin{bmatrix} A & B \\ C & D \end{bmatrix}\begin{bmatrix} V_R \\ I_R \end{bmatrix}$$
(c) matrix form of equations

Figure 37.3 *Transmission Line Two-Port Network*

Table 37.3 Per-Phase ABCD Constants for Transmission Lines

transmission line length	equivalent circuit	A	B	C	D
short <80 km	series impedance Fig. 37.4	1	Z	0	1
medium 80−240 km	nominal-T Fig. 37.5(a)	$1 + \frac{1}{2}YZ$	$Z\left(1 + \frac{1}{4}YZ\right)$	Y	$1 + \frac{1}{2}YZ$
medium 80−240 km	nominal-Π Fig. 37.5(b)	$1 + \frac{1}{2}YZ$	Z	$Y\left(1 + \frac{1}{4}YZ\right)$	$1 + \frac{1}{2}YZ$
long >240 km	distributed parameters Fig. 37.6	$\cosh \gamma l$	$Z_0 \sinh \gamma l$	$\dfrac{\sinh \gamma l}{Z_0}$	$\cosh \gamma l$

12. SHORT TRANSMISSION LINES

Short transmission lines are 60 Hz lines that are less than 80 km (50 mi) long.

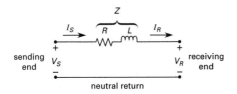

Figure 37.4 Short Transmission Line Model

13. MEDIUM-LENGTH TRANSMISSION LINES

Medium-length transmission lines are 60 Hz lines between 80 and 240 km (50 and 150 mi) long.

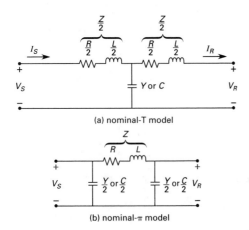

Figure 37.5 Medium-Length Transmission Line Models

14. LONG TRANSMISSION LINES

Long transmission lines are 60 Hz lines greater than 240 km (150 mi) long.

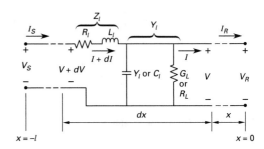

Figure 37.6 Long Transmission Line Model

15. REFLECTION COEFFICIENT

$$\text{VSWR} = \frac{V_{\max}}{V_{\min}} \qquad 37.56$$

$$\text{ISWR} = \frac{I_{\max}}{I_{\min}} \qquad 37.57$$

$$\Gamma_L = \frac{V_{\text{reflected}}}{V_{\text{incident}}} = \frac{Z_{\text{load}} - Z_0}{Z_{\text{load}} + Z_0} \qquad 37.58$$

$$\Gamma_L = \frac{I_{\text{reflected}}}{I_{\text{incident}}} = \frac{Z_0 - Z_{\text{load}}}{Z_0 + Z_{\text{load}}} \qquad 37.59$$

The fraction of incident power that is reflected back to the source from the load is Γ^2.

$$\Gamma = \frac{\text{SWR} - 1}{\text{SWR} + 1} \qquad 37.60$$

Power—Machinery

EERM Chapter 39
Rotating DC Machinery

Chapter, section, equation, figure, and table numbers correspond to EERM. For additional study material, go to the corresponding chapter and section number in EERM.

2. TORQUE AND POWER

$$T_{\text{ft-lbf}} = \frac{5252 P_{\text{horsepower}}}{n_{\text{rpm}}} \qquad 39.1$$

$$T_{\text{N·m}} = \frac{1000 P_{\text{kW}}}{\Omega} = \frac{9549 P_{\text{kW}}}{n_{\text{rpm}}} \qquad 39.2$$

Equation 39.3 is the general torque expression for a rotating machine with N coils of cross-sectional area A, each carrying current I through a magnetic field of strength B.

$$T = NBAI \cos \omega t \qquad 39.3$$

3. SERVICE FACTOR

$$\text{service factor} = \frac{\text{safe load}}{\text{nameplate load}} \qquad 39.4$$

6. REGULATION

$$\text{VR} = \frac{\begin{array}{c}\text{no-load voltage}\\ - \text{ full-load voltage}\end{array}}{\text{full-load voltage}} \times 100\% \qquad 39.7$$

$$\text{SR} = \frac{\begin{array}{c}\text{no-load speed}\\ - \text{ full-load speed}\end{array}}{\text{full-load speed}} \times 100\% \qquad 39.8$$

9. SERIES-WIRED DC MACHINES

$$V = E + I_a(R_a + R_f) \qquad 39.15$$

The speed, torque, and current are related by

$$\frac{T_1}{T_2} = \left(\frac{I_{a,1}}{I_{a,2}}\right)^2 \approx \frac{n_1}{n_2} \qquad 39.16$$

$$T = k_T' \Phi I_a = k_T I_a^2 \qquad 39.17$$

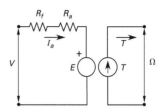

Figure 39.6 *Series-Wired DC Motor Equivalent Circuit*

10. SHUNT-WIRED DC MACHINES

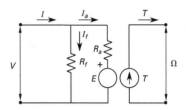

Figure 39.7 *Shunt-Wired DC Motor Equivalent Circuit*

$$V = E + I_a R_a = I_f R_f \qquad 39.20$$

$$I = I_a + I_f \qquad [\text{motor}] \qquad 39.21$$

$$I = I_a - I_f \qquad [\text{generator}] \qquad 39.22$$

$$n = n_{\text{nl}} - k_n T = \frac{V - I_a R_a}{k_E \Phi} \qquad 39.23$$

$$\frac{T_1}{T_2} = \frac{I_{a,1}}{I_{a,2}} \qquad 39.25$$

12. VOLTAGE-CURRENT CHARACTERISTICS FOR DC GENERATORS

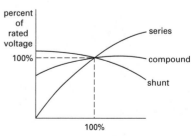

Figure 39.8 *DC Generator Voltage-Current Characteristics*

13. TORQUE CHARACTERISTICS FOR DC MOTORS

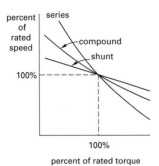

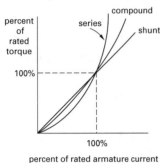

Figure 39.9 DC Motor Torque Characteristics

EERM Chapter 40
Rotating AC Machinery

Chapter, section, equation, figure, and table numbers correspond to EERM. For additional study material, go to the corresponding chapter and section number in EERM.

1. ROTATING MACHINES

It is common to refer to line-to-line voltage as the *terminal voltage*, V.

$$V_p = \begin{cases} V & \text{[delta-wired]} \\ \dfrac{V}{\sqrt{3}} & \text{[wye-wired]} \end{cases} \qquad 40.1$$

$$T_p = \frac{T_t}{3} \qquad 40.2$$

$$P_p = \frac{P_t}{3} \qquad 40.3$$

$$S_p = \frac{S_t}{3} \qquad 40.4$$

2. TORQUE AND POWER

$$T_{\text{ft-lbf}} = \frac{5252\,P_{\text{horsepower}}}{n_{\text{rpm}}} \qquad 40.5$$

$$T_{\text{N·m}} = \frac{1000\,P_{\text{kW}}}{\Omega} = \frac{9549\,P_{\text{kW}}}{n_{\text{rpm}}} \qquad 40.6$$

Equation 40.7 is the general torque expression for a rotating machine with N coils of cross-sectional area A each carrying current I through a magnetic field of strength B.

$$T = NBAI \cos \omega t \qquad 40.7$$

3. SERVICE FACTOR

$$\text{service factor} = \frac{\text{safe load}}{\text{nameplate load}} \qquad 40.8$$

6. REGULATION

$$\text{VR} = \frac{\begin{array}{c}\text{no-load voltage}\\ -\text{ full-load voltage}\end{array}}{\text{full-load voltage}} \times 100\% \qquad 40.11$$

$$\text{SR} = \frac{\begin{array}{c}\text{no-load speed}\\ -\text{ full-load speed}\end{array}}{\text{full-load speed}} \times 100\% \qquad 40.12$$

7. NO-LOAD CONDITIONS

The meaning of the term *no load* is different for generators and motors.

$$I = 0; \quad I_f \neq 0; \quad I_a = I_f \qquad \text{[generator]} \qquad 40.13$$

$$I \neq 0; \quad I_f = I; \quad I_a = 0 \qquad \text{[motor]} \qquad 40.14$$

8. PRODUCTION OF AC POTENTIAL

$$V = \frac{V_m}{\sqrt{2}} = \frac{\omega NAB}{\sqrt{2}} = \frac{p\Omega NAB}{2\sqrt{2}}$$

$$= \frac{\pi np NAB}{60\sqrt{2}} \qquad \text{[effective]} \qquad 40.16$$

$$n_s = \frac{120f}{p} = \frac{60\Omega}{2\pi} = \frac{60\omega}{\pi p} \qquad \begin{bmatrix}\text{synchronous}\\ \text{speed}\end{bmatrix} \qquad 40.17$$

$$f = \frac{1}{T} = \frac{\omega}{2\pi} = \frac{pn_s}{120} \qquad 40.18$$

12. SYNCHRONOUS MACHINE EQUIVALENT CIRCUIT

$$\mathbf{E} = \mathbf{V}_p + (R_a + jX_s)\mathbf{I}_a$$

$$\approx \mathbf{V}_p + jX_s\mathbf{I}_a \qquad \text{[alternator]} \qquad 40.21$$

$$\mathbf{V}_p = \mathbf{E} + (R_a + jX_s)\mathbf{I}_a$$

$$\approx \mathbf{E} + jX_s\mathbf{I}_s \qquad \text{[motor]} \qquad 40.22$$

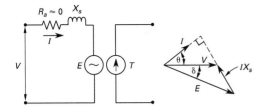

Figure 40.3 *Synchronous Motor Equivalent Circuit*

$$P_p = T_p \Omega = VI \cos\theta = \left(\frac{VE}{X_s}\right) \sin\delta \qquad 40.23$$

13. INDUCTION MOTORS

Slip in rpm is the difference between actual and synchronous speeds.

$$s = \frac{n_s - n}{n_s} = \frac{\Omega_s - \Omega}{\Omega_s} \qquad 40.26$$

14. INDUCTION MOTOR EQUIVALENT CIRCUIT

Using an adjusted voltage, V_{adj}, simplifies the model, as shown in Fig. 40.5(b). The relationship between the applied terminal voltage, V_1, and the adjusted voltage is

$$\mathbf{V}_{\text{adj}} = \mathbf{V}_1 - \mathbf{I}_{\text{nl}}(R_1 + jX_1) \qquad 40.27$$

$$V_{\text{adj}} \approx V_1 - I_{\text{nl}}\sqrt{R_1^2 + X_1^2} \qquad 40.28$$

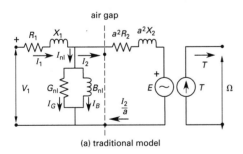

(a) traditional model

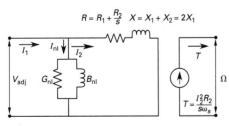

(b) simplified model ($a = 1$)

Figure 40.5 *Equivalent Circuits of an Induction Motor*

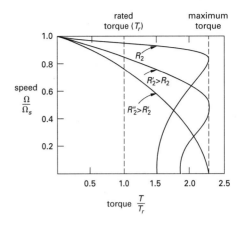

Figure 40.6 *Characteristic Curves for an Induction Motor*

19. POWER TRANSFER IN INDUCTION MOTORS

$$\text{input power} = V_1 I_1 \cos\theta \qquad 40.47$$

$$\text{stator copper losses} = I_1^2 R_1 \qquad 40.48$$

$$\text{rotor input power} = \frac{I_2^2 R_2}{s} \qquad 40.49$$

$$\text{rotor copper losses} = I_2^2 R_2 \qquad 40.50$$

$$\text{electrical power delivered} = I_2^2 R_2 \left(\frac{1-s}{s}\right) \qquad 40.51$$

$$\text{shaft output power} = T\Omega \qquad 40.52$$

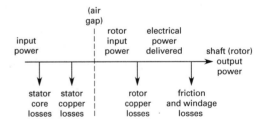

Figure 40.7 *Induction Motor Power Transfer*

Measurement and Instrumentation

EERM Chapter 42
Measurement and Instrumentation

Chapter, section, equation, figure, and table numbers correspond to EERM. For additional study material, go to the corresponding chapter and section number in EERM.

2. SIGNAL REPRESENTATION

$$V_{\text{ave}} = \left(\frac{2}{\pi}\right) V_p \qquad 42.7$$

$$V_{\text{rms}} = \left(\frac{1}{\sqrt{2}}\right) V_p \qquad 42.8$$

$$V_{\text{rms}} = 1.11 V_{\text{ave}} \qquad 42.9$$

3. MEASUREMENT CIRCUIT TYPES

When balanced, the relationship between the resistors is

$$\frac{R_x}{R_3} = \frac{R_2}{R_4} \qquad 42.10$$

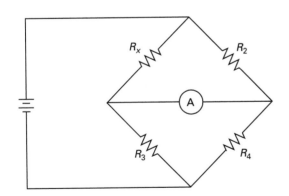

Figure 42.1 *Wheatstone Bridge*

5. DC VOLTMETERS

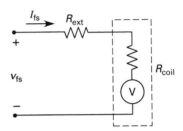

Figure 42.3 *DC Voltmeter*

$$\frac{1}{I_{\text{fs}}} = \frac{R_{\text{ext}} + R_{\text{coil}}}{V_{\text{fs}}} \qquad 42.11$$

6. DC AMMETERS

$$I_{\text{design}} = \frac{V_{\text{fs}}}{R_{\text{shunt}}} + I_{\text{fs}} \qquad 42.12$$

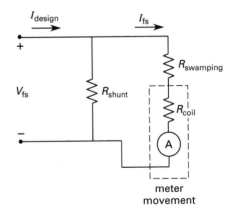

Figure 42.4 *DC Ammeter*

Electronics

Chapter, section, equation, figure, and table numbers correspond to EERM. For additional study material, go to the corresponding chapter and section number in EERM.

1. OVERVIEW

- *mass action law*

$$n_i^2 = np \qquad 43.1$$

2. SEMICONDUCTOR MATERIALS

- *density of electron-hole pairs in intrinsic materials*

$$n_i^2 = A_0 T^3 e^{-\frac{E_{G0}}{\kappa T}} \qquad 43.2$$

$$n_i^2 = N_c N_v e^{-\frac{E_G}{\kappa T}} \qquad 43.3$$

- *law of electrical neutrality*

$$N_A + n = N_D + p \qquad 43.5$$

- *concentration of electrons in a p-type material*

$$n = \frac{n_i^2}{N_A} \qquad 43.6$$

- *concentration of holes in an n-type material*

$$p = \frac{n_i^2}{N_D} \qquad 43.7$$

5. AMPLIFIERS

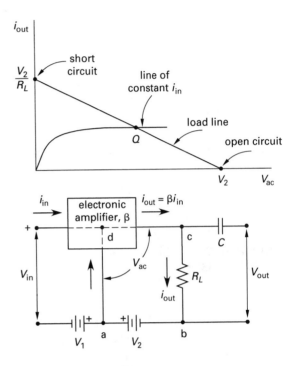

Figure 43.2 *General Amplifier*

7. LOAD LINE AND QUIESCENT POINT CONCEPT

Determination of the load line for a generic transistor amplifier is accomplished through the following steps.

step 1: For the configuration provided, label the *x*-axis on the *output characteristic curves* with the appropriate voltage. (For a BJT, this is V_{CE} or V_{CB}. For a FET, this is V_D.)

step 2: Label the *y*-axis as the output current. (For a BJT, this is I_C. For a FET, this is I_D.)

step 3: Redraw the circuit with all three terminals of the transistor open. Label the terminals. (For a BJT, these are base, emitter, and collector. For a FET, these are gate, source, and drain.) Label the current directions all pointing inward, toward the amplifier. (For a BJT, these are I_B, I_E, and I_C. For a FET, these are I_D and I_S.)

step 4: Perform KVL analysis in the output loop. (For a BJT, this is the collector loop. For a FET, this is the drain loop.) The transistor voltage determined is a point on the x-axis with the output current equal to zero. Plot the point.

step 5: Redraw the circuit with all three terminals of the transistor shorted. Label as in step 3.

step 6: Use Ohm's law, or another appropriate method, in the output loop to determine the current. (For the BJT, this is the collector current. For the FET, this is the drain current.) The transistor current determined is a point on the y-axis with the applicable voltage in step 1 equal to zero. Plot the point.

step 7: Draw a straight line between the two points. This is the DC load line.

An ideal *pn* junction, excluding the breakdown region, is governed by

$$I_{\text{pn}} = I_s \left(e^{\frac{qV_{\text{pn}}}{\kappa T}} - 1 \right) \qquad \textit{43.15}$$

9. DIODE PERFORMANCE CHARACTERISTICS

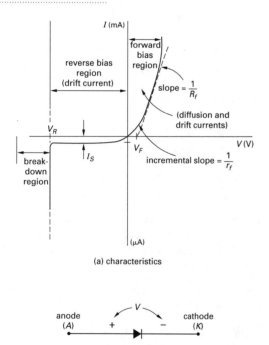

(a) characteristics

(b) symbol

Figure 43.9 Semiconductor Diode Characteristics and Symbol

For practical junctions (*diodes* or *rectifiers*),

$$I = I_s \left(e^{\frac{qV}{\eta\kappa T}} - 1 \right) = I_s \left(e^{\frac{V}{\eta V_T}} - 1 \right) \qquad \textit{43.16}$$

- *voltage equivalent of temperature*

$$V_T = \frac{\kappa T}{q} = \frac{D_p}{\mu_p} = \frac{D_n}{\mu_n} \qquad \textit{43.17}$$

The temperature has the effect of doubling the saturation current every 10°C.

$$\frac{I_{s2}}{I_{s1}} = (2)^{\frac{T_2 - T_1}{10°C}} \qquad \textit{43.18}$$

- *dynamic forward resistance*

$$R_f = r_f = \frac{\eta V_T}{I_D} \qquad \textit{43.19}$$

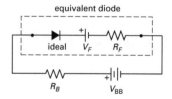

Figure 43.10 Diode Equivalent Circuit

10. DIODE LOAD LINE

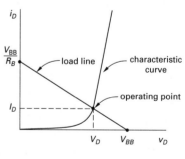

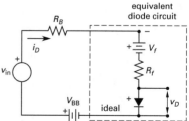

Figure 43.11 Diode Load Line

11. DIODE PIECEWISE LINEAR MODEL

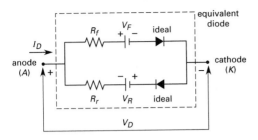

Figure 43.12 *Piecewise Linear Model*

16. PHOTODIODES AND LIGHT-EMITTING DIODES

- *emitted wavelength,* λ

$$\lambda = \frac{hc}{E_G} \qquad 43.26$$

17. SILICON-CONTROLLED RECTIFIERS

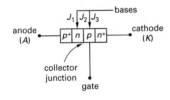

(a) conceptual construction

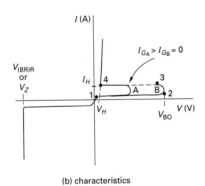

(b) characteristics

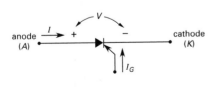

(c) symbol

Figure 43.17 *Silicon-Controlled Rectifier*

18. TRANSISTOR FUNDAMENTALS

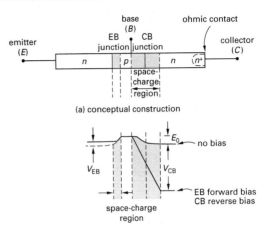

(a) conceptual construction

(b) electron energy level

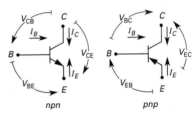

npn   *pnp*
(all currents shown in positive direction)
(all voltages shown for active region biasing)

(c) symbol

Figure 43.20 *Bipolar Junction Transistor*

19. BJT TRANSISTOR PERFORMANCE CHARACTERISTICS

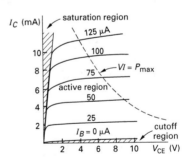

(a) common emitter

Figure 43.22 *BJT Output Characteristics*

20. BJT TRANSISTOR PARAMETERS

$$I_{\mathrm{E}} = I_C + I_B \qquad 43.28$$

$$\beta_{\mathrm{DC}} = \frac{I_C}{I_B} = \frac{\alpha_{\mathrm{DC}}}{1 - \alpha_{\mathrm{DC}}} \qquad 43.29$$

$$\alpha_{\mathrm{DC}} = \frac{I_C}{I_E} = \frac{\beta_{\mathrm{DC}}}{1 + \beta_{\mathrm{DC}}} \qquad 43.30$$

The difference between $\beta_{\rm ac}$ and $\beta_{\rm DC}$ is very small, and the two are not usually distinguished.

$$\beta_{\rm ac} = \frac{\Delta I_C}{\Delta I_B} = \frac{i_C}{i_B} \qquad 43.31$$

$$\alpha_{\rm ac} = \frac{\Delta I_C}{\Delta I_E} = \frac{i_C}{i_E} \qquad 43.32$$

$I_{\rm CBO}$ is the thermal current at the collector-base junction.

$$I_C = I_E - I_B \qquad 43.33$$

$$I_C = \alpha I_E - I_{\rm CBO} \approx \alpha I_E \qquad 43.34$$

- *equivalence of symbols*

$$\alpha_{\rm DC} = h_{\rm FB} \qquad 43.35$$

$$\alpha_{\rm ac} = h_{\rm fb} \qquad 43.36$$

$$\beta_{\rm DC} = h_{\rm FE} \qquad 43.37$$

$$\beta_{\rm ac} = h_{\rm fe} \qquad 43.38$$

23. BJT LOAD LINE

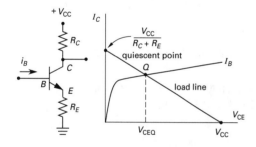

Figure 43.26 *Common Emitter Load Line and Quiescent Point*

24. AMPLIFIER GAIN AND POWER

$$A_V = \frac{\Delta V_{\rm out}}{\Delta V_{\rm in}} = \frac{v_{\rm out}}{v_{\rm in}} = \beta A_R \qquad 43.43$$

$$A_I = \frac{\Delta I_{\rm out}}{\Delta I_{\rm in}} = \frac{i_{\rm out}}{i_{\rm in}} = \beta \qquad 43.44$$

$$A_R = \frac{Z_{\rm out}}{Z_{\rm in}} = \frac{A_V}{\beta} \qquad 43.45$$

$$A_P = \frac{P_{\rm out}}{P_{\rm in}} = \beta^2 A_R = A_I A_V \qquad 43.46$$

26. EQUIVALENT CIRCUIT REPRESENTATION AND MODELS

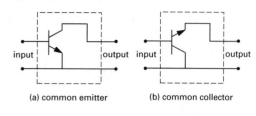

(a) common emitter   (b) common collector

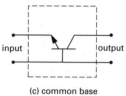

(c) common base

Figure 43.27 *Transistors as Two-Port Networks*

h_i = input impedance with output shorted (Ω)

h_r = reverse transfer voltage ratio with input open (dimensionless)

h_f = forward transfer current ratio with output shorted (dimensionless)

h_o = output admittance with input open (S)

Table 43.2 *Equivalent Circuit Parameters*

symbol	common emitter	common collector	common base
$h_{11}, h_{\rm ie}$	$h_{\rm ie}$	$h_{\rm ic}$	$\dfrac{h_{\rm ib}}{1 + h_{\rm fb}}$
$h_{12}, h_{\rm re}$	$h_{\rm re}$	$1 - h_{\rm rc}$	$\dfrac{h_{\rm ib}h_{\rm ob}}{1 + h_{\rm fb}} - h_{\rm rb}$
$h_{21}, h_{\rm fe}$	$h_{\rm fe}$	$-1 - h_{\rm fc}$	$\dfrac{-h_{\rm fb}}{1 + h_{\rm fb}}$
$h_{22}, h_{\rm oe}$	$h_{\rm oe}$	$h_{\rm oc}$	$\dfrac{h_{\rm ob}}{1 + h_{\rm fb}}$
$h_{11}, h_{\rm ib}$	$\dfrac{h_{\rm ie}}{1 + h_{\rm fe}}$	$\dfrac{-h_{\rm ic}}{h_{\rm fc}}$	$h_{\rm ib}$
$h_{12}, h_{\rm rb}$	$\dfrac{h_{\rm ie}h_{\rm oe}}{1 + h_{\rm fe}} - h_{\rm re}$	$h_{\rm rc} - \dfrac{h_{\rm ic}h_{\rm oc}}{h_{\rm fc}} - 1$	$h_{\rm rb}$
$h_{21}, h_{\rm fb}$	$\dfrac{-h_{\rm fe}}{1 + h_{\rm fe}}$	$\dfrac{-1 + h_{\rm fc}}{h_{\rm fc}}$	$h_{\rm fb}$
$h_{22}, h_{\rm ob}$	$\dfrac{h_{\rm oe}}{1 + h_{\rm fe}}$	$\dfrac{-h_{\rm oc}}{h_{\rm fc}}$	$h_{\rm ob}$
$h_{11}, h_{\rm ic}$	$h_{\rm ie}$	$h_{\rm ic}$	$\dfrac{h_{\rm ib}}{1 + h_{\rm fb}}$
$h_{12}, h_{\rm rc}$	$1 - h_{\rm re}$	$h_{\rm rc}$	1
$h_{21}, h_{\rm fc}$	$-1 - h_{\rm fe}$	$h_{\rm fc}$	$\dfrac{-1}{1 + h_{\rm fb}}$
$h_{22}, h_{\rm oc}$	$h_{\rm oe}$	$h_{\rm oc}$	$\dfrac{h_{\rm ob}}{1 + h_{\rm fb}}$

27. APPROXIMATE TRANSISTOR MODELS

The values of h_r and h_o are very small. The simplified
models of Table 43.5 are obtained by ignoring these two
parameters.

Table 43.5 BJT Simplified Equivalent Circuits

common connection	equivalent circuit	network equations
CE	CE	common emitter[a] $v_{be} = h_{ie}i_b \approx 0.7$ V $i_c = h_{fe}i_b$
CC	CC	common collector $v_{bc} = h_{ic}i_b$ $i_e = h_{fc}i_b$
CB	CB	common base[a] $v_{eb} = h_{ib}i_e \approx 0.7$ V $i_c = h_{fb}i_e$

[a]Germanium transistors are *pnp* types. $|v_{be}| = |v_{eb}| = 0.3$ for germanium.

34. JFET CHARACTERISTICS

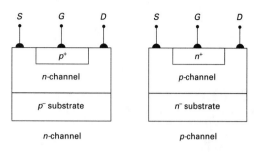

(a) conceptual construction

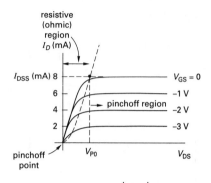

n-channel

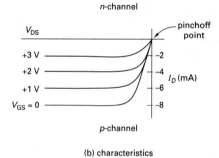

p-channel

(b) characteristics

(c) symbol

Figure 43.32 *Junction Field-Effect Transistor*

The term "pinchoff voltage" and the symbol V_P are ambiguous, as the actual pinchoff voltage in a circuit depends on the gate-source voltage, V_{GS}. When V_{GS} is zero, the pinchoff voltage is represented unambiguously by V_{P0}.

$$V_P = V_{P0} + V_{GS} \qquad 43.60$$

- *Shockley's equation*

$$I_D = I_{DSS}\left(1 - \frac{V_{GS}}{V_P}\right)^2 \qquad 43.61$$

- *transconductance, g_m*

$$g_m = \frac{\Delta I_D}{\Delta V_{GS}} = \frac{i_D}{v_{GS}}$$

$$= \left(\frac{-2I_{DSS}}{V_P}\right)\left(1 - \frac{V_{GS}}{V_P}\right) = g_{mo}\left(1 - \frac{V_{GS}}{V_P}\right)$$

$$\approx \frac{A_V}{R_{out}} \qquad 43.62$$

The drain-source resistance can be obtained from the slope of the V_{GS} characteristic in Fig. 43.32(b).

$$r_d = r_{DS} = \frac{\Delta V_{DS}}{\Delta I_D} = \frac{v_{DS}}{i_D} \qquad 43.63$$

35. JFET BIASING

- *load line equation*

$$V_{DD} = I_S(R_D + R_S) + V_{DS}$$

$$= I_D(R_D + R_S) + V_{DS} \qquad 43.64$$

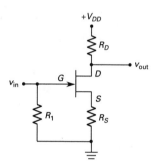

Figure 43.34 *Self-Biasing JFET Circuit*

At the quiescent point, $V_{in} = 0$. From Kirchhoff's voltage law, around the input loop,

$$V_{GS} = -I_S R_S = -I_D R_S \qquad 43.65$$

36. FET MODELS

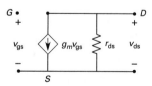

Figure 43.35 *FET Equivalent Circuit*

37. MOSFET CHARACTERISTICS

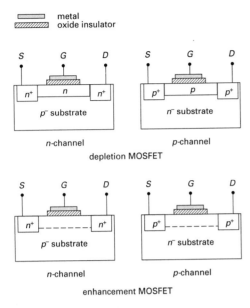

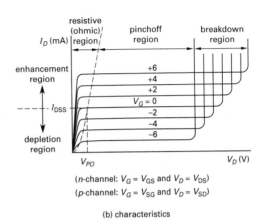

(b) characteristics

(a) conceptual construction

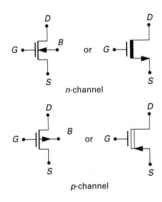

(c) depletion MOSFET symbols

(d) enhancement MOSFET symbols

Figure 43.37 *Metal-Oxide Semiconductor Field-Effect Transistor*

38. MOSFET BIASING

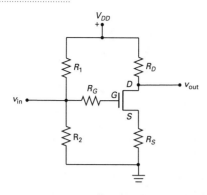

Figure 43.38 *Typical MOSFET Biasing Circuit*

Since the gate current is zero, the voltage divider is unloaded.

$$V_G = V_{DD} \left(\frac{R_2}{R_1 + R_2} \right) = V_{GS} + I_S R_S \qquad \textbf{\textit{43.66}}$$

The load line equation is found from Kirchhoff's voltage law and is the same as for the JFET.

$$V_{DD} = I_S(R_D + R_S) + V_{DS} \qquad \textbf{\textit{43.67}}$$

EERM Chapter 44
Amplifiers

Chapter, section, equation, figure, and table numbers correspond to EERM. For additional study material, go to the corresponding chapter and section number in EERM.

1. FUNDAMENTALS

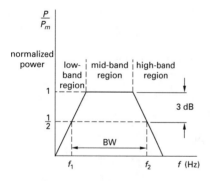

Figure 44.1 *Bandwidth*

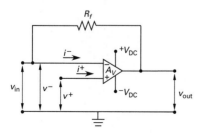

Figure 44.2 *Operational Amplifier Symbology*

$$v_{\text{out}} = A_V \left(v^+ - v^- \right) \qquad \textbf{\textit{44.3}}$$

- *input signal range with the distortion restriction*

$$\left| v^+ - v^- \right| < \frac{V_{\text{DC}} - 3 \text{ V}}{A_V} \qquad \textbf{\textit{44.4}}$$

In practice, both the *differential-mode signal*, v_{dm}, and the *common-mode signal*, v_{cm}, are amplified.

$$v_{\text{dm}} = v^+ - v^- \qquad \textbf{\textit{44.5}}$$

$$v_{\text{cm}} = \tfrac{1}{2} \left(v^+ + v^- \right) \qquad \textbf{\textit{44.6}}$$

- *common-mode rejection ratio*

$$\text{CMRR} = \left| \frac{A_{\text{dm}}}{A_{\text{cm}}} \right| \qquad \textbf{\textit{44.7}}$$

$$v_{\text{out}} = A_{\text{dm}} v_{\text{dm}} \left(1 + \left(\frac{1}{\text{CMRR}} \right) \left(\frac{v_{\text{cm}}}{v_{\text{dm}}} \right) \right) \qquad \textbf{\textit{44.8}}$$

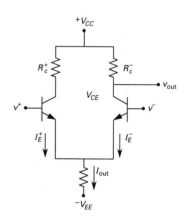

Figure 44.4 *Differential Amplifier*

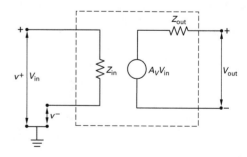

Figure 44.6 *Operational Amplifier Equivalent Circuit*

2. IDEAL OPERATIONAL AMPLIFIERS

An *ideal operational amplifier* exhibits the following properties.

- $Z_{\text{in}} = \infty$
- $Z_{\text{out}} = 0$
- $A_V = \infty$
- $\text{BW} = \infty$

The assumptions regarding the properties of the ideal op amp result in the following practical results during analysis.

- The current to each input is zero.
- The voltage between the two input terminals is zero.
- The op amp is operating in the linear range.

The voltage difference of zero between the two terminals is called a *virtual short circuit*, or, because the positive terminal is often grounded, a *virtual ground*.

3. OPERATIONAL AMPLIFIER LIMITS

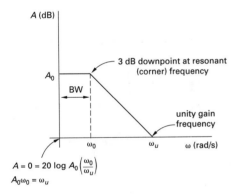

Figure 44.9 *Operational Amplifier Frequency Response*

- *unity gain*

$$|A| = A_0\left(\frac{\omega_0}{\omega_u}\right) = 1 \qquad 44.9$$

- *slew rate*

$$S_R = \left.\frac{dv}{dt}\right|_{\max} \approx \frac{I_{\max}}{C} \qquad 44.10$$

4. AMPLIFIER NOISE

- *thermal noise, P_n* (also called *noise power*)

$$P_n = \frac{V_{\text{noise,rms}}^2}{4R} = \kappa TB \qquad 44.11$$

$$\text{SNR} = \frac{S}{N} = 10\log\left(\frac{P_s}{P_n}\right) = 20\log\left(\frac{V_s}{V_n}\right) \qquad 44.12$$

EERM Chapter 45
Pulse Circuits: Waveform Shaping and Logic

Chapter, section, equation, figure, and table numbers correspond to EERM. For additional study material, go to the corresponding chapter and section number in EERM.

2. CLAMPING CIRCUITS

Analysis of a limiter circuit yields Eqs. 45.1 and 45.2.

$$v_{\text{out}} = \frac{R(V_{\text{ref}} - V_F)}{R + R_f} + \frac{R_f V_{\text{in}}}{R_f + R} \quad [V_{\text{in}} < V_{\text{ref}} - V_F]$$

$$45.1$$

$$V_{\text{out}} = V_{\text{in}} \quad [V_{\text{in}} > V_{\text{ref}} - V_F] \qquad 45.2$$

Equation 45.1 represents the forward-biased case. Equation 45.2 represents the reverse-biased case, ignoring the small reverse saturation current.

Analysis of a clipper circuit yields Eqs. 45.3 and 45.4.

$$v_{\text{out}} = \frac{R_f V_{\text{in}}}{R_f + R} + \frac{R(V_{\text{ref}} + V_F)}{R + R_f} \quad [V_{\text{in}} > V_{\text{ref}} + V_F]$$

$$45.3$$

$$V_{\text{out}} = V_{\text{in}} \quad [V_{\text{in}} < V_{\text{ref}} + V_F] \qquad 45.4$$

The approximate transfer equation for a precision diode is

$$V_{\text{out}} = V_{\text{in}} - \frac{V_F}{A_V} \qquad 45.5$$

4. ZENER VOLTAGE REGULATOR CIRCUIT: PRACTICAL

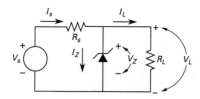

(a) regulator circuit

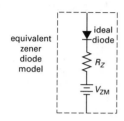

(b) regulator circuit with equivalent zener model

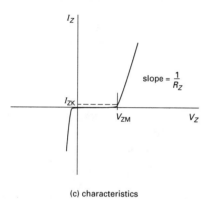

(c) characteristics

Figure 45.6 *Equivalent Zener Diode Voltage Regulator*

The defining equations for the equivalent circuit are

$$V_{\text{in}} - V_L = (I_L + I_Z)R_s \qquad 45.12$$

$$V_L = V_{\text{ZM}} + I_Z R_Z \qquad 45.13$$

Equation 45.14 is used to determine the source resistance with the input voltage at its minimum value and the load current at its maximum value.

$$V_L = \frac{V_{\text{ZM}}R_s + V_{\text{in}}R_Z}{R_s + R_Z} - I_L\left(\frac{R_sR_Z}{R_s + R_Z}\right) \qquad 45.14$$

$$I_{Z,\text{max}} = \frac{V_{\text{ZM}}R_s + V_{\text{in,max}}R_Z}{R_Z(R_s + R_Z)} - \frac{V_{\text{ZM}}}{R_Z}$$

$$= \frac{V_{\text{in,max}}}{R_s} - \frac{V_{\text{ZM}}R_s + V_{\text{in,max}}R_Z}{R_s(R_s + R_Z)} \qquad 45.15$$

The power requirements for the diode and the supply resistor are

$$P_D = I_{Z,\text{max}}V_{\text{ZM}} + I_{Z,\text{max}}^2 R_Z \qquad 45.16$$

$$P_{R_s} = I_{Z,\text{max}}^2 R_s \qquad 45.17$$

6. TRANSISTOR SWITCH FUNDAMENTALS

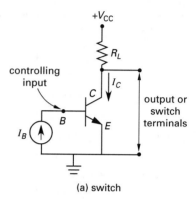

(a) switch

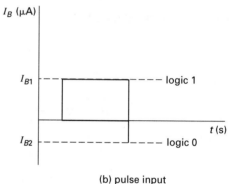

(b) pulse input

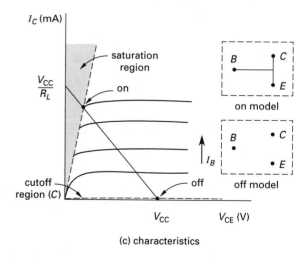

(c) characteristics

Figure 45.8 *Transistor Switch*

When the transistor switch is in the on state, that is, saturated,

$$I_{C,\text{sat}} = \frac{V_{\text{CC}} - V_{\text{CE,sat}}}{R_L} \qquad 45.24$$

The collector-emitter voltage, V_{CE}, is typically 0.1 V for germanium diodes and 0.2 V for silicon diodes and can be ignored for first-order calculations—hence the short-circuit model of Fig. 45.8(c). The value of the load resistance must be determined so that when in saturation the condition of Eq. 45.25 is satisfied.

$$I_{B1} \geq \frac{I_{C,\text{sat}}}{h_{\text{FE}}} \approx \frac{V_{CC}}{h_{\text{FE}} R_L} \qquad 45.25$$

7. JFET SWITCHES

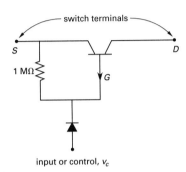

Figure 45.9 JFET Switch

8. CMOS SWITCHES

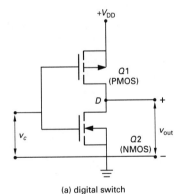

(a) digital switch

(b) characteristics

Figure 45.10 CMOS Inverter

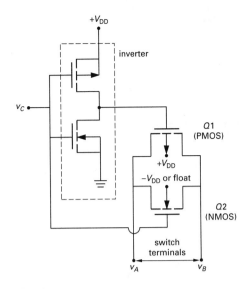

Figure 45.11 CMOS Switch

11. LOGIC GATES

Table 45.1 Logic Gates

inputs		not	and	or	nand	nor	exclusive or
A	B	$-A$ or $\overline{A}$	AB	$A+B$	$\overline{AB}$	$\overline{A+B}$	$A \oplus B$
0	0	1	0	0	1	1	0
0	1	1	0	1	1	0	1
1	0	0	0	1	1	0	1
1	1	0	1	1	0	0	0

12. BOOLEAN ALGEBRA

- *commutative*

$$A + B = B + A$$
$$A \cdot B = B \cdot A$$

- *associative*

$$A + (B + C) = (A + B) + C$$
$$xA \cdot (B \cdot C) = (A \cdot B) \cdot C$$

- *distributive*

$$A \cdot (B + C) = (A \cdot B) + (A \cdot C)$$
$$A + (B \cdot C) = (A + B) \cdot (A + C)$$

- *absorptive*

$$A + (A \cdot B) = A$$
$$A \cdot (A + B) = A$$

- *De Morgan's theorems*

$$\overline{A + B} = \overline{A} \cdot \overline{B}$$

$$\overline{A \cdot B} = \overline{A} + \overline{B}$$

13. LOGIC CIRCUIT FAN-OUT

Fan-out is the number of parallel loads that can be driven from one output node of a logic circuit.

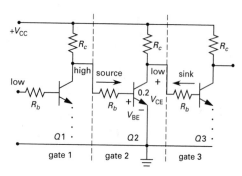

Figure 45.13 *Fan-Out Unit Loads*

- *source fan-out*

$$N + \frac{R_b}{R_c} \leq \frac{V_{\text{CC}} - V_{\text{BE}} - I_{\text{CBO}} R_c}{I_{B,\min} R_c} \qquad 45.33$$

- *sink fan-out*

$$N \leq \frac{I_{C,\text{sat}} R_c + V_{\text{CE,sat}} - V_{\text{CC}}}{I_{\text{CBO}} R_c} \qquad 45.34$$

Table 45.2 *Logic Family Data*

parameter	RTL	DTL	HTL	TTL	ECL	MOS	CMOS
basic gate[a]	NOR	NAND	NAND	NAND	OR-NOR	NAND	NOR OR NAND
fan-out[b]	5	8	10	10	25	20	> 50
propagation delay[c]	12	30	90	10	0.1	10	0.1

[a] Positive logic is assumed when determining the basic gate.
[b] Worst-case condition.
[c] Approximate values in nanoseconds.

Computers

EERM Chapter 46
Computer Hardware Fundamentals

> Chapter, section, equation, figure, and table numbers correspond to EERM. For additional study material, go to the corresponding chapter and section number in EERM.

4. MICROPROCESSORS

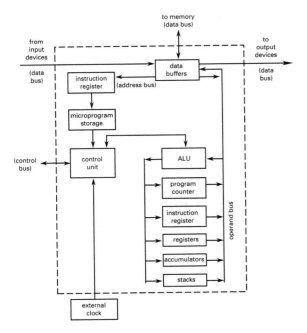

Figure 46.2 *Microprocessor Architecture*

EERM Chapter 47
Computer Software Fundamentals

> Chapter, section, equation, figure, and table numbers correspond to EERM. For additional study material, go to the corresponding chapter and section number in EERM.

1. CHARACTER CODING

American Standard Code for Information Interchange (ASCII) is a seven-bit code permitting 128 (2^7) different combinations.

The *Extended Binary Coded Decimal Interchange Code* (EBCDIC) is in widespread use in mainframe computers. It uses eight bits (a byte) for each character, allowing a maximum of 256 (2^8) different characters.

EERM Chapter 49
Digital Logic

> Chapter, section, equation, figure, and table numbers correspond to EERM. For additional study material, go to the corresponding chapter and section number in EERM.

3. FUNDAMENTAL LOGIC OPERATIONS

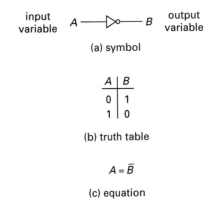

Figure 49.2 *NOT Logic*

Table 49.2 Logic Operators

operator	symbol	truth table	equation

AND

A	B	C
0	0	0
0	1	0
1	0	0
1	1	1

$A \cdot B = C$

OR

A	B	C
0	0	0
0	1	1
1	0	1
1	1	1

$A + B = C$

XOR

A	B	C
0	0	0
0	1	1
1	0	1
1	1	0

$A \oplus B = C$

NAND

A	B	C
0	0	1
0	1	1
1	0	1
1	1	0

$\overline{A \cdot B} = C$

NOR

A	B	C
0	0	1
0	1	0
1	0	0
1	1	0

$\overline{A + B} = C$

XNOR or coincidence

A	B	C
0	0	1
0	1	0
1	0	0
1	1	1

$A \odot B = C$
or
$\otimes$

4. MINTERMS AND MAXTERMS

Table 49.3 Minterms and Maxterms

decimal row number	binary input combinations ABC	minterm (product term)	maxterm (sum term)
0	000	$m_0 = \overline{A}\,\overline{B}\,\overline{C}$	$M_0 = A + B + C$
1	001	$m_1 = \overline{A}\,\overline{B}C$	$M_1 = A + B + \overline{C}$
2	010	$m_2 = \overline{A}B\overline{C}$	$M_2 = A + \overline{B} + C$
3	011	$m_3 = \overline{A}BC$	$M_3 = A + \overline{B} + \overline{C}$
4	100	$m_4 = A\overline{B}\,\overline{C}$	$M_4 = \overline{A} + B + C$
5	101	$m_5 = A\overline{B}C$	$M_5 = \overline{A} + B + \overline{C}$
6	110	$m_6 = AB\overline{C}$	$M_6 = \overline{A} + \overline{B} + C$
7	111	$m_7 = ABC$	$M_7 = \overline{A} + \overline{B} + \overline{C}$

5. CANONICAL REPRESENTATION OF LOGIC FUNCTIONS

- *minterm form* (of the three-variable function in Table 49.3)

$$F(A,B,C) = \sum_{i=0}^{7} m_i \qquad 49.4$$

- *canonical sum-of-product form* (SOP)

$$
\begin{aligned}
F(A,B,C) &= m_0 + m_1 + m_2 + m_3 + m_4 + m_5 \\
&\quad + m_6 + m_7 \\
&= \overline{A}\,\overline{B}\,\overline{C} + \overline{A}\,\overline{B}C + \overline{A}B\overline{C} + \overline{A}BC \\
&\quad + A\overline{B}\,\overline{C} + A\overline{B}C + AB\overline{C} \\
&\quad + ABC \qquad 49.5
\end{aligned}
$$

- *maxterm form* (of the three-variable function in Table 49.3)

$$F(A,B,C) = \prod_{i=0}^{7} M_i \qquad 49.6$$

- *canonical product-of-sum form* (POS)

$$
\begin{aligned}
F(A,B,C) &= M_0 M_1 M_2 M_3 M_4 M_5 M_6 M_7 \\
&= (A + B + C)\left(A + B + \overline{C}\right) \\
&\quad \times \left(A + \overline{B} + C\right)\left(A + \overline{B} + \overline{C}\right) \\
&\quad \times \left(\overline{A} + B + C\right)\left(\overline{A} + B + \overline{C}\right) \\
&\quad \times \left(\overline{A} + \overline{B} + C\right)\left(\overline{A} + \overline{B} + \overline{C}\right) \qquad 49.7
\end{aligned}
$$

EERM Chapter 50
Logic Network Design

Chapter, section, equation, figure, and table numbers correspond to EERM. For additional study material, go to the corresponding chapter and section number in EERM.

4. KARNAUGH MAP FUNDAMENTALS

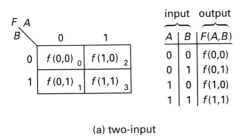

F A B	0	1
0	$f(0,0)_0$	$f(1,0)_2$
1	$f(0,1)_1$	$f(1,1)_3$

input		output
A	B	$F(A,B)$
0	0	$f(0,0)$
0	1	$f(0,1)$
1	0	$f(1,0)$
1	1	$f(1,1)$

(a) two-input

F AB CD	00	01	11	10	adjacency ordering
00	$f(0,0,0,0)_0$	$f(0,1,0,0)_4$	$f(1,1,0,0)_{12}$	$f(1,0,0,0)_8$	
01	$f(0,0,0,1)_1$	$f(0,1,0,1)_5$	$f(1,1,0,1)_{13}$	$f(1,0,0,1)_9$	
11	$f(0,0,1,1)_3$	$f(0,1,1,1)_7$	$f(1,1,1,1)_{15}$	$f(1,0,1,1)_{11}$	
10	$f(0,0,1,0)_2$	$f(0,1,1,0)_6$	$f(1,1,1,0)_{14}$	$f(1,0,1,0)_{10}$	

adjacency ordering

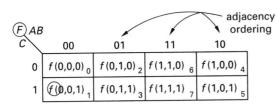

F AB C	00	01	11	10	adjacency ordering
0	$f(0,0,0)_0$	$f(0,1,0)_2$	$f(1,1,0)_6$	$f(1,0,0)_4$	
1	$f(0,0,1)_1$	$f(0,1,1)_3$	$f(1,1,1)_7$	$f(1,0,1)_5$	

input			output
A	B	C	$F(A,B,C)$
0	0	0	$f(0,0,0)$
0	0	1	$f(0,0,1)$
0	1	0	$f(0,1,0)$
0	1	1	$f(0,1,1)$
1	0	0	$f(1,0,0)$
1	0	1	$f(1,0,1)$
1	1	0	$f(1,1,0)$
1	1	1	$f(1,1,1)$

(b) three-input

input				output
A	B	C	D	$F(A,B,C,D)$
0	0	0	0	$f(0,0,0,0)$
0	0	0	1	$f(0,0,0,1)$
0	0	1	0	$f(0,0,1,0)$
0	0	1	1	$f(0,0,1,1)$
0	1	0	0	$f(0,1,0,0)$
0	1	0	1	$f(0,1,0,1)$
0	1	1	0	$f(0,1,1,0)$
0	1	1	1	$f(0,1,1,1)$
1	0	0	0	$f(1,0,0,0)$
1	0	0	1	$f(1,0,0,1)$
1	0	1	0	$f(1,0,1,0)$
1	0	1	1	$f(1,0,1,1)$
1	1	0	0	$f(1,1,0,0)$
1	1	0	1	$f(1,1,0,1)$
1	1	1	0	$f(1,1,1,0)$
1	1	1	1	$f(1,1,1,1)$

(c) four-input

Figure 50.3 *Karnaugh Maps and Associated Truth Tables*

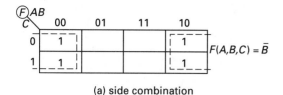

(a) side combination

$$F(A,B,C) = \bar{B}$$

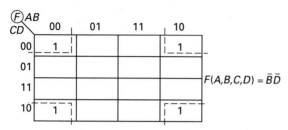

(b) corner combination

$$F(A,B,C,D) = \bar{B}\bar{D}$$

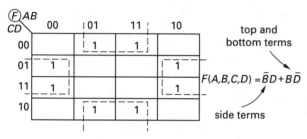

top and bottom terms

$$F(A,B,C,D) = \bar{B}D + B\bar{D}$$

side terms

(c) side and top/bottom combination

Figure 50.4 *Typical Adjacent Minterm Groupings*

The general rules for combining or grouping minterms (or maxterms) are as follows.

- Each one-square (zero-square) must be accounted for in at least one grouping.

- Any grouping should be as large as possible.

- Groupings are made in terms of 1, 2, 4, 8, ..., 2^n minterms (maxterms).

- A one-square grouping should not be used if it can be combined with another one-square. A two-square grouping should not be used if it can be combined into a four-square, and so on.

- Groupings should be used in such a manner as to minimize the total number of groups.

6. VEITCH DIAGRAMS

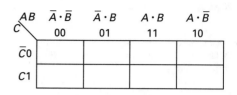

(a) three-input named variables

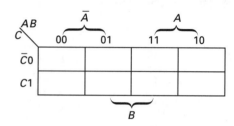

(b) three-input Veitch diagram

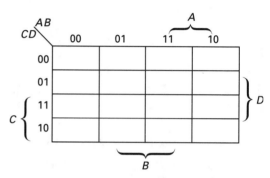

(c) four-input Veitch diagram

Figure 50.5 *Veitch Diagram Fundamentals*

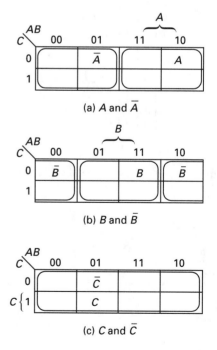

(a) A and $\bar{A}$

(b) B and $\bar{B}$

(c) C and $\bar{C}$

Figure 50.6 *Three-Input K-V Map: Minterms*

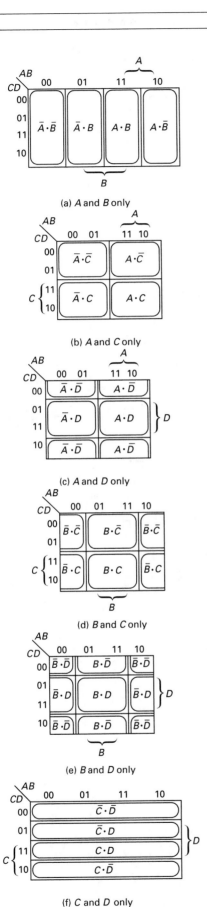

(a) A and B only

(b) A and C only

(c) A and D only

(d) B and C only

(e) B and D only

(f) C and D only

Figure 50.7 *Four-Input K-V Map: Minterms*

7. NAND/NOR FUNCTION REPRESENTATION

step 1: For a NAND only expression, group the minterms (output value one) in a map to form the sum-of-product (SOP) expression. For a NOR only expression, group the maxterms (output value zero) in a map to form the product-of-sum (POS) expression.

step 2: Complement the expression from step 1.

step 3: Apply De Morgan's law to the expression in step 2.

step 4: Complement the result.

8. PROGRAMMABLE LOGIC ARRAYS

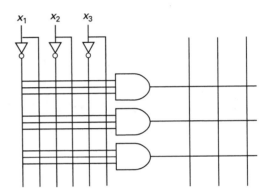

(a) unprogrammed

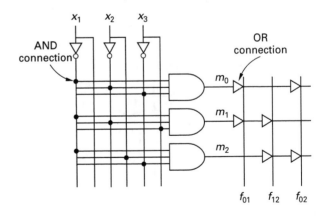

(b) programmed

Figure 50.9 *Programmable Logic Array*

9. ENCODERS AND DECODERS

(a) priority encoder block diagram

input				output	
x_3	x_2	x_1	x_0	Q_1	Q_0
0	0	0	0	d	d
0	0	0	1	0	0
0	0	1	0	0	1
0	1	0	0	1	0
1	0	0	0	1	1

(b) priority encoder truth table

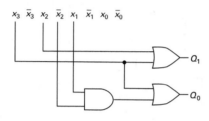

(c) AND-OR realization

Figure 50.10 *Encoder*

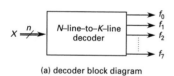

(a) decoder block diagram

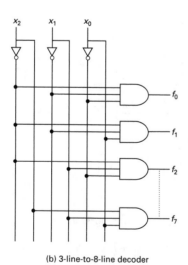

(b) 3-line-to-8-line decoder

Figure 50.11 *Decoder*

10. DEMULTIPLEXERS

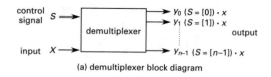

(a) demultiplexer block diagram

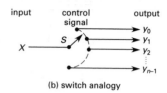

(b) switch analogy

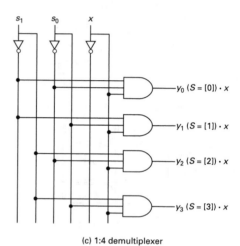

(c) 1:4 demultiplexer

Figure 50.12 *Demultiplexer*

11. MULTIPLEXERS

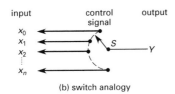

(a) multiplexer block diagram

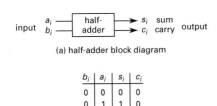

(b) switch analogy

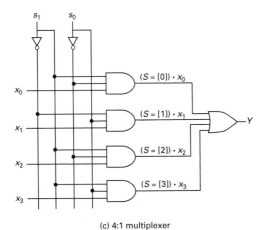

(c) 4:1 multiplexer

Figure 50.13 *Multiplexer*

12. ARITHMETIC LOGIC DEVICES

Binary addition is accomplished with the following rules.

- $0 + 0 = \text{sum } 0 \text{ carry } 0$
- $0 + 1 = \text{sum } 1 \text{ carry } 0$
- $1 + 0 = \text{sum } 1 \text{ carry } 0$
- $1 + 1 = \text{sum } 0 \text{ carry } 1$

(a) half-adder block diagram

b_i	a_i	s_i	c_i
0	0	0	0
0	1	1	0
1	0	1	0
1	1	0	1

(b) half-adder truth table

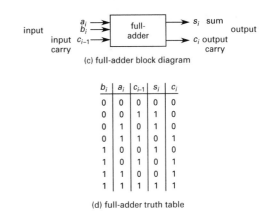

(c) full-adder block diagram

b_i	a_i	c_{i-1}	s_i	c_i
0	0	0	0	0
0	0	1	1	0
0	1	0	1	0
0	1	1	0	1
1	0	0	1	0
1	0	1	0	1
1	1	0	0	1
1	1	1	1	1

(d) full-adder truth table

Figure 50.14 *Adders*

EERM Chapter 51
Synchronous Sequential Networks

> Chapter, section, equation, figure, and table numbers correspond to EERM. For additional study material, go to the corresponding chapter and section number in EERM.

1. FUNDAMENTALS

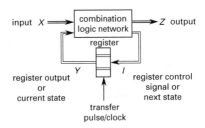

Figure 51.1 *General Synchronous Sequential Network*

The general steps in designing a synchronous sequential network are as follows.

step 1: Describe the network specifications.

step 2: Determine the state table required.

step 3: Minimize the state table.

step 4: Make the state assignments; that is, encode the information in binary form in order to transform the state table into a transition table.

step 5: Realize the network.

2. CLOCKS AND TIMING

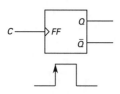

(a) rising edge-triggered
flip-flop

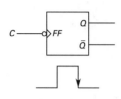

(b) falling edge-triggered
flip-flop

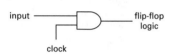

(c) clock logic realization

Figure 51.2 *Flip-Flop Clocks*

3. *S-R* FLIP-FLOPS

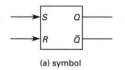

(a) symbol

S	R	Q	Q⁺	remarks
0	0	0	0	hold
0	0	1	1	hold
0	1	0	0	reset
0	1	1	0	reset
1	0	0	1	set
1	0	1	1	set
1	1	0	indeterminate	disallowed condition
1	1	1	indeterminate	disallowed condition

(b) state table

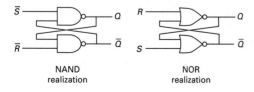

| NAND realization | NOR realization |

(c) logic realization

Figure 51.3 *S-R Flip-Flop*

- *next-state equation* (also called the *characteristic equation* or *transition equation*) *for the S-R flip-flop*

$$Q^+ = S + \overline{R}Q \qquad 51.1$$

4. *J-K* FLIP-FLOPS

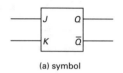

(a) symbol

J	K	Q	Q⁺	remarks
0	0	0	0	hold
0	0	1	1	hold
0	1	0	0	reset
0	1	1	0	reset
1	0	0	1	set
1	0	1	1	set
1	1	0	1	toggle
1	1	1	0	toggle

(b) state table

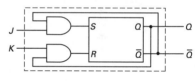

(c) logic realization

Figure 51.4 *J-K Flip-Flop*

- *next-state equation for the J-K flip-flop*

$$Q^+ = J\overline{Q} + \overline{K}Q \qquad 51.2$$

5. *T* FLIP-FLOPS

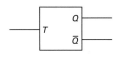

(a) symbol

T	Q	Q⁺	remarks
0	0	0	hold
0	1	1	hold
1	0	1	toggle
1	1	0	toggle

(b) state table

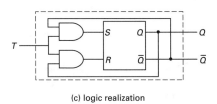

(c) logic realization

Figure 51.5 *T Flip-Flop*

- *next-state equation for the T flip-flop*

$$Q^+ = T \oplus Q = T\overline{Q} + \overline{T}Q \qquad 51.3$$

6. *D* FLIP-FLOPS

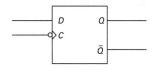

(a) symbol

D	Q	Q⁺	remarks
0	0	0	The output follows
0	1	0	the input, *D*,
1	0	1	regardless of the
1	1	1	present state, *Q*.

(b) state table

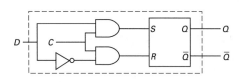

(c) logic realization
(edge trigger circuitry not shown)

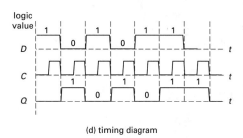

(d) timing diagram

Figure 51.6 *D Flip-Flop*

- *next-state equation for the D flip-flop*

$$Q^+ = D \qquad 51.4$$

7. FLIP-FLOP TRANSITION TABLES

Table 51.1 *Flip-Flop Transition Table*

present state Q	next state Q^+	S	R	J	K	T	D	
0	⟶ 0	0	X	0	X	0	0	$Q^+ = S + \overline{R} \cdot Q$
0	⟶ 1	1	0	1	X	1	1	$Q^+ = J \cdot \overline{Q} + \overline{K} \cdot Q$
1	⟶ 0	0	1	X	1	1	0	$Q^+ = T \oplus Q$
1	⟶ 1	X	0	X	0	0	1	$Q^+ = D$

EERM Chapter 54
Operating Systems, Networking Systems, and Standards

Chapter, section, equation, figure, and table numbers correspond to EERM. For additional study material, go to the corresponding chapter and section number in EERM.

3. INTERNET FUNDAMENTALS

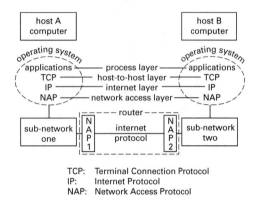

TCP: Terminal Connection Protocol
IP: Internet Protocol
NAP: Network Access Protocol

Figure 54.2 *Transmission Control Protocol/*
Internet Protocol Overview

4. NETWORK ORGANIZATION

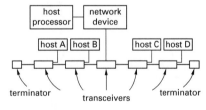

(a) logical view of bus topology

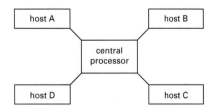

(b) physical view of star topology

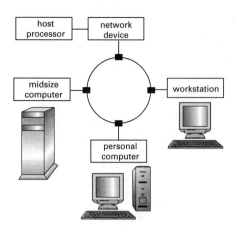

(c) logical view of ring topology

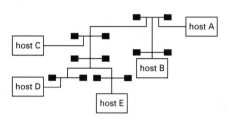

(d) logical view of tree topology

Figure 54.3 *Network Topologies*

EERM Chapter 55
Digital Systems

Chapter, section, equation, figure, and table numbers correspond to EERM. For additional study material, go to the corresponding chapter and section number in EERM.

2. INTERFACE FUNDAMENTALS

The process of matching the flow of information between portions of a digital system is called the *interface problem.* The general solution to interface issues is shown in Fig. 55.2.

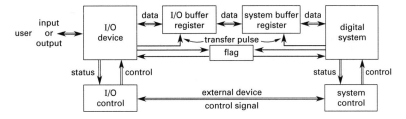

Figure 55.2 *Hardware Interface Overview*

3. DATA CONVERSION BASICS

The value of a positive n-bit binary number N is

$$N = b_{n-1}2^{n-1} + b_{n-2}2^{n-2} + \cdots + b_2 2^2$$
$$+ b_1 2^1 + b_0 2^0 \qquad \textit{55.1}$$

When coded in binary, N is

$$N = b_{n-1}b_{n-2}\ldots b_2 b_1 b_0 \qquad \textit{55.2}$$

The digit b_{n-1} is the most significant bit (MSB) while b_0 is the least significant bit (LSB). The maximum value that can be represented by the code in Eq. 55.2 is

$$\text{maximum value} = (2)\,(\text{MSB}) - \text{LSB}$$
$$= 2^n - 1 \qquad \textit{55.3}$$

Communications

EERM Chapter 56
High-Frequency Transmission

Chapter, section, equation, figure, and table numbers correspond to EERM. For additional study material, go to the corresponding chapter and section number in EERM.

2. HIGH-FREQUENCY TRANSMISSION LINES

At high frequencies, approximately 1 MHz and higher, wavelengths are shorter and even a few feet of line are treated as a long transmission line.

$$Z_l = jX_{L,l} = j\omega L_l \qquad 56.1$$

$$Y_l = jB_{C,l}$$
$$= \frac{j}{X_{C,l}} = -\frac{1}{jX_{C,l}} = j\omega C_l \qquad 56.2$$

- *propagation constant*, γ (in units of radians per meter (mile))

$$\gamma = \sqrt{Y_l Z_l}$$
$$= \sqrt{(j\omega C_l)(j\omega L_l)}$$
$$= j\omega\sqrt{L_l C_l}$$
$$= j\beta \qquad 56.3$$

The term β in Eq. 56.3 is the phase constant.

- *phase velocity*, v_{phase}

$$v_{phase} = \frac{1}{\sqrt{LC}} \qquad 56.4$$

- *wavelength*

$$\lambda = \frac{2\pi}{\beta} = \frac{1}{f\sqrt{LC}} \qquad 56.5$$

- *characteristic impedance*, Z_0

$$Z_0 = \sqrt{\frac{Z_l}{Y_l}} = \sqrt{\frac{j\omega L_l}{j\omega C_l}} = \sqrt{\frac{L}{C}} \qquad 56.6$$

$$Z_{in} = Z_0\left(\frac{(1+\Gamma)\cos\beta l + j(1-\Gamma)\sin\beta l}{(1-\Gamma)\cos\beta l + j(1+\Gamma)\sin\beta l}\right) \qquad 56.10$$

$$Z_{in} = Z_0\left(\frac{Z_{load}\cos\beta l + jZ_0\sin\beta l}{Z_0\cos\beta l + jZ_{load}\sin\beta l}\right) \qquad 56.11$$

$$VSWR = \frac{|V^+| + |V^-|}{|V^+| - |V^-|} = \frac{1+|\Gamma|}{1-|\Gamma|} \qquad 56.12$$

$$Z_{max} = Z_0(VSWR) \qquad 56.13$$

$$Z_{min} = \frac{Z_0}{VSWR} \qquad 56.14$$

3. SMITH CHART FUNDAMENTALS

A *Smith chart* is a special polar diagram with constant reflection coefficient circles, constant standing wave ratio circles, constant resistance circles, constant reactance arcs (portions of a circle), and radius lines representing constant line-angle loci. The Smith chart is essentially a polar representation of the reflection coefficient in terms of the normalized resistance and reactance. The normalization occurs with respect to the characteristic impedance, Z_0.

$$z = \frac{Z}{Z_0} = \frac{R+jX}{Z_0} = \frac{1+\Gamma}{1-\Gamma} \qquad 56.15$$

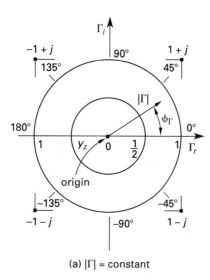

(a) $|\Gamma|$ = constant

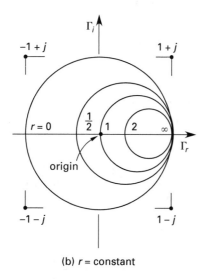

(b) r = constant

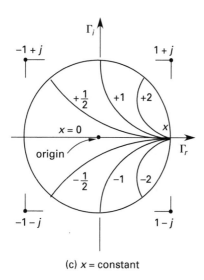

(c) x = constant

Figure 56.1 *Smith Chart Components*

Table 56.1 *Smith Chart Electrical Conditions*

electrical condition	reflection coefficient (Γ)	normalized resistance (r)	normalized reactance (x)
open circuit	$1\angle 0°$	∞ (arbitrary)	arbitrary (∞)
short circuit	$1\angle 180°$	0	0
pure reactance	$1\angle \pm 90°$	0	± 1
matched line (pure resistance)	0	1	0

EERM Chapter 57
Antenna Theory

> Chapter, section, equation, figure, and table numbers correspond to EERM. For additional study material, go to the corresponding chapter and section number in EERM.

3. FREE-SPACE WAVE CHARACTERISTICS

$$\lambda = \frac{2\pi}{\beta} = \frac{2\pi}{\omega\sqrt{\mu_0\varepsilon_0}} = \frac{c}{f} \qquad 57.3$$

- *velocity of an electromagnetic wave in free space*

$$\mathrm{v}_0 = \frac{1}{\sqrt{\mu_0\varepsilon_0}} \approx c \approx 3 \times 10^8 \ \mathrm{m/s} \qquad 57.4$$

- *characteristic impedance of free space*

$$Z_0 = \sqrt{\frac{\mu_0}{\varepsilon_0}} = 120\pi \ \Omega = 377 \ \Omega$$

$$= 4\pi\mathrm{v}_0 \times 10^{-7} \ \Omega \qquad 57.5$$

4. POYNTING'S VECTOR AND POWER

$$\mathbf{S} = \varepsilon c E^2 \mathbf{a} = \mathbf{E} \times \mathbf{H} \quad [\text{in W/m}^2] \qquad 57.6$$

$$P_{\mathrm{ave}} = \tfrac{1}{2}\mathrm{Re}\left(\mathbf{E} \times \mathbf{H}\right) = \frac{E^2}{2Z_0} \quad [\text{in W/m}^2] \qquad 57.7$$

7. REFLECTION

At the *critical incident angle*, θ_c, an optically transparent surface becomes totally reflecting.

$$\sin\theta_c = \frac{1}{n} \qquad 57.8$$

The critical angle depends on the *absolute index of refraction, n.*

$$n = \sqrt{\mu_r \varepsilon_r} \qquad 57.9$$

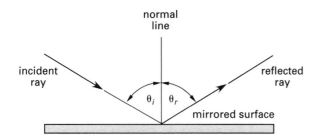

Figure 57.4 *Reflection from a Surface*

8. REFRACTION

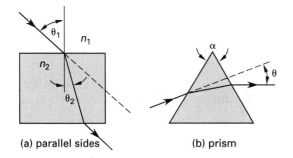

Figure 57.5 *Refraction*

$$n_{\text{relative}} = \frac{n_2}{n_1} = \frac{\sin \theta_1}{\sin \theta_2} \qquad 57.10$$

$$n_{\text{relative}} = \frac{\sin \frac{1}{2}(\alpha + \theta)}{\sin \frac{1}{2}\alpha} \qquad 57.11$$

9. DIFFRACTION AND SCATTERING

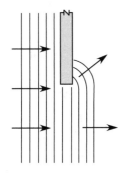

Figure 57.6 *Diffraction*

10. FREE-SPACE BASIC TRANSMISSION LOSS

The gain, G, and loss, L, in a communication system are usually given in decibels (dB).

$$L_{b0} = 20 \log \left(\frac{4\pi r}{\lambda} \right) \qquad 57.14$$

The term $(4\pi r/\lambda)^2$ is called the *path loss*.

11. ELEMENTAL ANTENNA

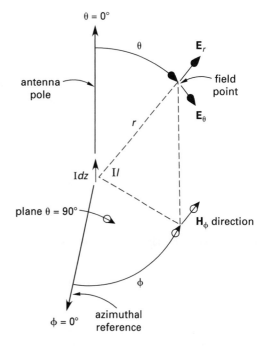

Figure 57.7 *Elemental Antenna*

The expression for the magnetic field strength or intensity, $\mathbf{H}$, of an elemental antenna of length $dz = l$, using spherical coordinates, is

$$H_\phi = \left(\frac{Il}{4\pi} \right) \sin \theta \left(j\frac{\beta}{r} + \frac{1}{r^2} \right)$$
$$\times e^{-j\beta r} \quad [\text{in A/m}] \qquad 57.15$$

The components of the electric field, $\mathbf{E}$, are found using Maxwell's equation relationship.

$$\mathbf{\nabla} \times \mathbf{H} = \text{curl } \mathbf{H} = \frac{\partial \mathbf{D}}{\partial t} = j\omega \varepsilon_0 \mathbf{E} \qquad 57.16$$

- *radial component of the electric field*

$$E_r = \left(\frac{Il}{2\pi} \right) \cos \theta \left(\frac{1}{cr^2} + \frac{1}{j\beta r^3} \right)$$
$$\times e^{-j\beta r} \quad [\text{in V/m}] \qquad 57.17$$

- *polar component*

$$E_\theta = \left(\frac{Il}{4\pi}\right) \sin\theta \left(\frac{j\beta}{r} + \frac{1}{r^2} + \frac{1}{j\beta r^3}\right)$$
$$\times e^{-j\beta r} \quad [\text{in V/m}] \qquad 57.18$$

12. ELEMENTAL ANTENNA: RADIATED POWER

- *radiated power density*

$$p_r = \left(\frac{Il}{4\pi}\right)^2 \sqrt{\frac{\mu_0}{\varepsilon_0}} \sin^2\theta$$
$$\times \left(\frac{j\beta}{r}\right)^2 e^{-2j\beta r} \quad [\text{in W/m}^2] \qquad 57.19$$

- *average radiated power in an elemental antenna of length l*

$$P_r = \left(\frac{(Il)^2}{6\pi}\right)\sqrt{\frac{\mu_0}{\varepsilon_0}}\beta^2 \quad [\text{in W/m}^2] \qquad 57.20$$

13. ELEMENTAL ANTENNA: DIRECTIVITY

The *directivity* is the value of the directive gain of an antenna in the direction of its maximum value.

$$D_g = \frac{P_{\text{isotropic}}}{P_r} \qquad 57.21$$

$$P_{r,\text{max}} = \left(\frac{Il}{4\pi}\right)^2 \sqrt{\frac{\mu_0}{\varepsilon_0}} \left(\frac{j\beta}{r}\right)^2$$
$$\times e^{-2j\beta r} \quad [\text{in W/m}^2] \qquad 57.22$$

An isotropic antenna radiates power equally in all directions, forming a sphere with a surface area of $4\pi r^2$.

$$P_{\text{isotropic}} = \left(\frac{(Il)^2}{4\pi}\right)\sqrt{\frac{\mu_0}{\varepsilon_0}}\beta^2 \quad [\text{in W/m}^2] \qquad 57.23$$

- *directivity*

$$D_g = \frac{P_{\text{isotropic}}}{P_r} = 1.5 \qquad 57.24$$

- *directivity of the elemental antenna* (in terms of decibels)

$$D_G = 10\log D_g = 1.76 \text{ dB} \qquad 57.25$$

14. ELEMENTAL ANTENNA: RADIATION RESISTANCE

- *radiation resistance*, R_{rad}

$$R_{\text{rad}} = \frac{P_r}{I_{\text{in}}^2} \qquad 57.26$$

$$R_{\text{rad}} = \left(\frac{l^2\beta^2}{6\pi}\right)\sqrt{\frac{\mu_0}{\varepsilon_0}} \qquad 57.27$$

15. ELEMENTAL ANTENNA: OHMIC RESISTANCE

$$R_{\text{ohmic}} = \frac{R_l}{1 - \left(1 - \dfrac{\delta}{r}\right)^2} \qquad 57.28$$

16. ELEMENTAL ANTENNA: EFFICIENCY

$$\eta = \frac{R_{\text{rad}}}{R_{\text{ohmic}} + R_{\text{rad}}} \times 100\% \qquad 57.29$$

17. ELEMENTAL ANTENNA: GAIN

$$G = \eta D_g \qquad 57.30$$

EERM Chapter 58
Communication Links

> Chapter, section, equation, figure, and table numbers correspond to EERM. For additional study material, go to the corresponding chapter and section number in EERM.

2. POWER DISTRIBUTION

The power density radiated from the transmitting antenna available at the receiving antenna is

$$p_r = \left(\frac{D_T}{4\pi r^2}\right)P_T \quad [\text{in W/m}^2] \qquad 58.1$$

In free space, the power density is

$$p_r = \frac{E^2}{Z_0} = \frac{E^2}{\sqrt{\dfrac{\mu_0}{\varepsilon_0}}} = \frac{E^2}{120\pi} \quad [\text{in W/m}^2] \qquad 58.2$$

The power lost in the radiation resistance is reradiated. This power is considered *scattered*.

$$P_{\text{scat}} = \frac{120\pi p_r l^2 R_{\text{rad}}}{\left(R_{\text{in}} + R_{\text{amp}}\right)^2 + \left(X_{\text{in}} + X_{\text{amp}}\right)^2} \qquad 58.4$$

The power lost in the ohmic resistance of the antenna is

$$P_{\text{loss}} = \frac{120\pi p_r l^2 R_{\text{ohmic}}}{\left(R_{\text{in}} + R_{\text{amp}}\right)^2 + \left(X_{\text{in}} + X_{\text{amp}}\right)^2} \qquad 58.5$$

The power utilized by the amplifier is

$$P_{\text{amp}} = \frac{V_{\text{rec}}^2 R_{\text{amp}}}{\left(R_{\text{in}} + R_{\text{amp}}\right)^2 + \left(X_{\text{in}} + X_{\text{amp}}\right)^2}$$

$$= \frac{120\pi p_r l^2 R_{\text{amp}}}{\left(R_{\text{in}} + R_{\text{amp}}\right)^2 + \left(X_{\text{in}} + X_{\text{amp}}\right)^2} \qquad 58.6$$

3. ANTENNA APERTURES

An antenna aperture is an opening through which radio waves can pass.

$$A_{\text{scat}} = \frac{120\pi l^2 R_{\text{rad}}}{\left(R_{\text{in}} + R_{\text{amp}}\right)^2 + \left(X_{\text{in}} + X_{\text{amp}}\right)^2} \qquad 58.7$$

$$A_{\text{loss}} = \frac{120\pi l^2 R_{\text{ohmic}}}{\left(R_{\text{in}} + R_{\text{amp}}\right)^2 + \left(X_{\text{in}} + X_{\text{amp}}\right)^2} \qquad 58.8$$

$$A_{\text{eff}} = \frac{120\pi l^2 R_{\text{amp}}}{\left(R_{\text{in}} + R_{\text{amp}}\right)^2 + \left(X_{\text{in}} + X_{\text{amp}}\right)^2} \qquad 58.9$$

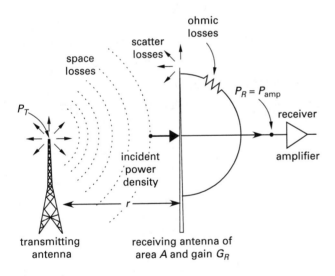

Note: The power density is sometimes called the flux density.

Figure 58.1 *Communication Link*

The relationships in Eqs. 58.10 through 58.13 are valid regardless of the antenna type.

$$P_{\text{scat}} = A_{\text{scat}} p_r \qquad 58.10$$

$$P_{\text{loss}} = A_{\text{loss}} p_r \qquad 58.11$$

$$P_{\text{amp}} = A_{\text{eff}} p_r \qquad 58.12$$

$$p_{r,\,\text{total}} = \left(A_{\text{scat}} + A_{\text{loss}} + A_{\text{eff}}\right) p_r$$

$$= A_{\text{total}} p_r \qquad 58.13$$

4. DIRECTIVITY

$$D_g = \left(\frac{4\pi}{\lambda^2}\right) A_{\text{eff}} \qquad 58.14$$

5. BEAM SOLID ANGLE Ω

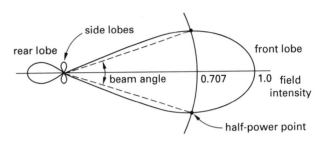

Figure 58.2 *Beam Angle*

$$P_r = p_{r,\text{max}} \Omega r^2 \qquad 58.15$$

$$D_g = \frac{4\pi}{\Omega} \qquad 58.16$$

6. COMMUNICATION LINK TRANSMISSION

$$P_R = \left(\frac{D_T A_R}{4\pi r^2}\right) P_T \qquad 58.17$$

$$P_R = \left(\frac{A_R A_T}{\lambda^2 r^2}\right) P_T \qquad 58.18$$

$$P_R = \left(\frac{\lambda}{4\pi r}\right)^2 D_R D_T P_T \qquad 58.19$$

The total efficiency relates the gain to the directivity.

$$G = \eta D \qquad 58.20$$

The receiving antenna power must be written in terms of the gain when the efficiencies are significant, that is, low, giving

- *Friss transmission formula*

$$P_R = \left(\frac{\lambda}{4\pi r}\right)^2 G_R G_T P_T \qquad 58.21$$

In word form, the formula is

$$\frac{\text{power}}{\text{received}} = \frac{\text{EIRP} \times \text{gain of receiving antenna}}{\text{path loss}} \qquad 58.22$$

$$P_R = \text{EIRP} + G_R - L_p \quad \text{[in dBW]} \qquad 58.23$$

$$\text{EIRP} = 10 \log\left(G_T P_T\right) \quad \text{[in dBW]} \qquad 58.24$$

$$G_R = 10 \log\left(\eta\left(\frac{4\pi A_R}{\lambda^2}\right)\right) \quad \text{[in dBW]} \qquad 58.25$$

$$L_p = 20 \log\left(\frac{4\pi r}{\lambda}\right) \quad \text{[in dBW]} \qquad 58.26$$

EERM Chapter 59
Signal Formats

> Chapter, section, equation, figure, and table numbers correspond to EERM. For additional study material, go to the corresponding chapter and section number in EERM.

1. FUNDAMENTALS

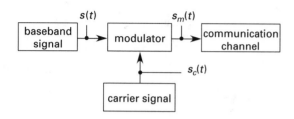

(a) transmission

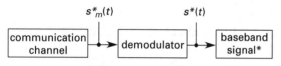

*Signal distortion possible. Original signal may be corrupted and require correction.

(b) reception

Figure 59.2 *Baseband Signal*

A carrier signal can be represented by the general equation

$$s_c(t) = A\cos(\omega_c t + \theta) = A\cos(2\pi f_c t + \theta) \qquad 59.1$$

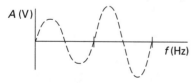

(a) amplitude modulation (AM)

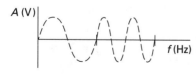

(b) frequency modulation (FM)

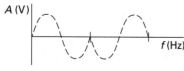

(c) phase modulation (PM)

Figure 59.3 *Modulation Types*

3. AMPLITUDE MODULATION

In amplitude modulation (AM), the angle, θ, of Eq. 59.1 is constant and usually assumed to be equal to zero.

$$S_{\text{mod}}(\omega) = \tfrac{1}{2}S(\omega - \omega_c) + \tfrac{1}{2}S(\omega + \omega_c) \qquad 59.3$$

When the information signal is a single frequency, $s(t) = a\cos\omega_{\text{mod}}t$,

$$s_{\text{mod}}(t) = A(m\cos\omega_{\text{mod}}t + 1)\cos\omega_c t \qquad 59.5$$

The dimensionless scaling factor in Eq. 59.5, m, is called the amplitude *modulation index, index of modulation,* or *modulation factor,* and is defined mathematically as

$$m_{\text{AM}} = \frac{ka}{A} = \frac{\Delta A}{A} \qquad 59.6$$

7. AMPLITUDE MODULATION: BANDWIDTH AND POWER

Amplitude modulation is a frequency translation. The allowable change in frequency of the modulated signal is

$$\text{BW} \le \Delta f \le 2(\text{BW}) \qquad 59.8$$

The lower limit corresponds to the single-sideband signal, the upper limit to the double-sideband signal.

9. FREQUENCY MODULATION

Frequency modulation (FM) changes the carrier wave frequency in proportion to the instantaneous value of the modulating wave.

$$s_{\text{mod}}(t) = A\cos(\omega_c t + m_{\text{FM}}\sin\omega_{\text{mod}}t) \qquad 59.13$$

The dimensionless ratio factor in Eq. 59.13 is called the *frequency modulation index, index of modulation,* or *modulation factor,* and is defined as

$$m_{\text{FM}} = \frac{\Delta\omega}{\omega_{\text{mod}}} \qquad 59.14$$

$$\Delta\omega = ak_f \qquad 59.15$$

The term a is the amplitude of the modulating signal, that is, the baseband or information signal. The *frequency modulator constant,* k_f, has units of radians per second per volt.

15. INTERSYMBOL INTERFERENCE

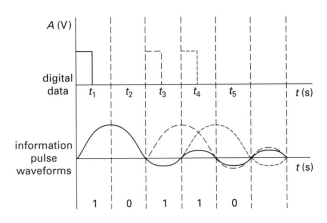

Figure 59.5 *Zero Intersymbol Interference*

19. DIGITAL MODULATION: BANDWIDTH AND POWER

- *Shannon bandwidth* (used to compare modulation schemes)

$$\text{BW} = \frac{D}{2T} \qquad 59.20$$

The variable D is called the *dimensionality* of the signal set. That is, it is the number of orthonormal signals within a given mapping (coding) scheme. The time duration over which the signals are defined is T.

The power is the product of the binary digit rate, R_s, in digits per second, and the bit energy, E_{bit}.

$$P = \frac{E_{\text{ave}}}{T} = E_{\text{bit}} R_s \qquad 59.21$$

- *bit energy*

$$E_{\text{bit}} = \frac{E_{\text{ave}}}{\log_2 M} \qquad 59.22$$

- *digit rate*

$$R_s = \frac{\log_2 M}{T} \qquad 59.23$$

The term M is the number of waveforms upon which the binary digits are mapped and is given by $M = 2^n$ where n is the number of binary digits available.

Assuming *white Gaussian noise* with a power spectral density of $N_0/2$, the *signal-to-noise ratio* (SNR) is

$$\text{SNR} = \frac{S}{N} = \frac{P}{N_0(\text{BW})} = \left(\frac{E_{\text{bit}}}{N_0}\right)\left(\frac{R_s}{\text{BW}}\right) \qquad 59.24$$

20. DIGITAL ERROR PROBABILITY

The performance of a given digital modulation scheme can be measured by the *symbol error probability*, $\mathcal{P}(e)$, which is the total probability that a given waveform is detected incorrectly. The probability of corruption for a single bit is called the *bit error probability*, $\mathcal{P}_b(e)$, also called the *bit error rate* (BER). Since each symbol carries $\log_2 M$ bits, a signal symbol error affects at a minimum, one bit, and at a maximum, $\log_2 M$ bits. The overall relationship is

$$\frac{\mathcal{P}(e)}{\log_2 M} \le \mathcal{P}_{\text{bit}}(e) \le \mathcal{P}(e) \qquad 59.25$$

EERM Chapter 61 Communications Systems and Channels

> Chapter, section, equation, figure, and table numbers correspond to EERM. For additional study material, go to the corresponding chapter and section number in EERM.

1. SYSTEMS AND CHANNELS OVERVIEW

The capacity, C, of a digital channel in bits per second (bps) is determined by the signal-to-noise ratio and the bandwidth.

$$C = (\text{BW})\log_2\left(1 + \frac{S}{N}\right)$$
$$= (\text{BW})\log_2\left(1 + \frac{S}{N_0(\text{BW})}\right) \qquad 61.1$$

- *Shannon limit* (the value of the channel capacity as the bandwidth approaches infinity)

$$C = 1.44\left(\frac{S}{N_0}\right) \qquad 61.2$$

Considering the curvature of the Earth, the radio horizon, d, in kilometers, depends on the antenna height, h, in meters, as shown in Eq. 61.3.

$$d = 4.1\sqrt{h} \qquad 61.3$$

Control Systems

EERM Chapter 63
Control Systems

Chapter, section, equation, figure, and table numbers correspond to EERM. For additional study material, go to the corresponding chapter and section number in EERM.

1. TYPES OF RESPONSE

Natural response (also known as *initial condition response*, *homogeneous response*, and *unforced response*) is the manner in which a system behaves when energy is applied and then subsequently removed.

Forced response is the behavior of a system that is acted upon by a force that is applied periodically.

2. GRAPHICAL SOLUTION

When the system equation is a homogeneous second-order linear differential equation with constant coefficients in the form of Eq. 63.2, the natural time response can be determined from Fig. 63.1.

$$x'' + 2\zeta\omega x' + \omega^2 x = 0 \qquad 63.2$$

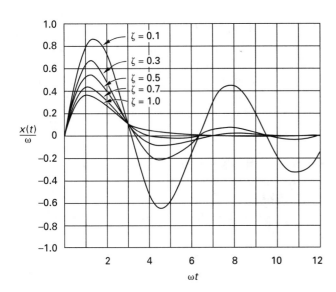

Figure 63.1 *Natural Response*

When the system equation is a second-order linear differential equation with constant coefficients and the forcing function is a step of height h (as in Eq. 63.3), the time response can be determined from Fig. 63.2.

$$x'' + 2\zeta\omega x' + \omega^2 x = \omega^2 h \qquad 63.3$$

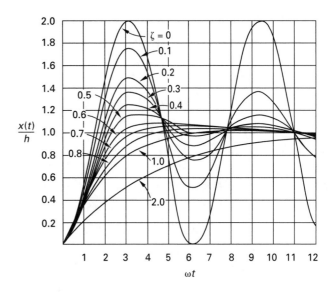

Figure 63.2 *Response to a Unit Step*

$$\omega_d = \text{damped frequency} = \omega\sqrt{1-\zeta^2} \qquad 63.4$$

$$t_r = \text{rise time} = \frac{\pi - \arccos\zeta}{\omega_d} \qquad 63.5$$

$$t_p = \text{peak time} = \frac{\pi}{\omega_d} \qquad 63.6$$

$$M_p = \text{peak gain (fraction overshoot)}$$

$$= \exp\left(\frac{-\pi\zeta}{\sqrt{1-\zeta^2}}\right) \qquad 63.7$$

$$t_s = \text{settling time}$$

$$= \frac{3.91}{\zeta\omega_n} \quad [\text{2\% criterion}] \qquad 63.8$$

$$= \frac{3.00}{\zeta\omega_n} \quad [\text{5\% criterion}] \qquad 63.9$$

$$\tau = \text{time constant}$$

$$= \frac{1}{\zeta\omega_n} \qquad 63.10$$

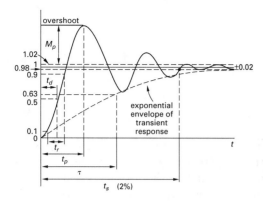

Figure 63.3 *Second-Order Step Time Response Parameters*

4. FEEDBACK THEORY

- *error transfer function (error gain)*, $E(s)$

$$E(s) = \mathcal{L}(e(t)) = V_i(s) \pm V_f(s)$$
$$= V_i(s) \pm H(s)V_o(s) \qquad \textbf{63.12}$$

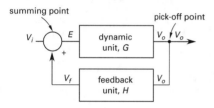

Figure 63.4 *Feedback System*

$$\frac{E(s)}{V_i(s)} = \frac{1}{1 + G(s)H(s)} \quad \text{[negative feedback]} \quad \textbf{63.13}$$

$$= \frac{1}{1 - G(s)H(s)} \quad \text{[positive feedback]} \quad \textbf{63.14}$$

Forward gain or *direct gain*, $G(s)$, is normally a complex operator that changes both the magnitude and the phase of the error.

$$V_o(s) = G(s)E(s) \qquad \textbf{63.15}$$

The transfer function of the feedback unit is the *reverse transfer function (feedback transfer function, feedback gain)*, $H(s)$, which can be a simple magnitude-changing scalar or a phase-shifting function.

$$V_f(s) = H(s)V_o(s) \qquad \textbf{63.16}$$

- *feedback ratio (primary feedback ratio)*, $V_f(s)/V_i(s)$

$$\frac{V_f(s)}{V_i(s)} = \frac{G(s)H(s)}{1 + G(s)H(s)} \quad \text{[negative feedback]} \quad \textbf{63.17}$$

$$= \frac{G(s)H(s)}{1 - G(s)H(s)} \quad \text{[positive feedback]} \quad \textbf{63.18}$$

The *loop transfer function (loop gain* or *open-loop transfer function)* is the gain after going around the loop one time, $\pm G(s)H(s)$.

The *overall transfer function (closed-loop transfer function, control ratio, system function, closed-loop gain)*, $G_{\text{loop}}(s)$, is the overall transfer function of the feedback system. The quantity $1 + G(s)H(s) = 0$ is the *characteristic equation*.

$$G_{\text{loop}}(s) = \frac{V_o(s)}{V_i(s)}$$

$$= \frac{G(s)}{1 + G(s)H(s)} \quad \text{[negative feedback]} \quad \textbf{63.19}$$

$$= \frac{G(s)}{1 - G(s)H(s)} \quad \text{[positive feedback]} \quad \textbf{63.20}$$

5. SENSITIVITY

The sensitivity of the loop transfer function with respect to the forward transfer function is

$$S_{G(s)}^{G_{\text{loop}}(s)} = \left(\frac{\Delta G_{\text{loop}}(s)}{\Delta G(s)} \right) \left(\frac{G(s)}{G_{\text{loop}}(s)} \right)$$

$$= \frac{1}{1 + G(s)H(s)} \quad \left[\begin{array}{c} \text{negative} \\ \text{feedback} \end{array} \right] \quad \textbf{63.22}$$

$$= \frac{1}{1 - G(s)H(s)} \quad \left[\begin{array}{c} \text{positive} \\ \text{feedback} \end{array} \right] \quad \textbf{63.23}$$

9. INITIAL AND FINAL VALUES

$$\lim_{t \to 0+} p(t) = \lim_{s \to \infty} (sP(s)) \quad \text{[initial value]} \quad \textbf{63.26}$$

$$\lim_{t \to \infty} p(t) = \lim_{s \to 0} (sP(s)) \quad \text{[final value]} \quad \textbf{63.27}$$

11. POLES AND ZEROS

A *pole* is a value of s that makes a function, $P(s)$, infinite. Specifically, a pole makes the denominator of $P(s)$ zero. Poles off the real axis always occur in conjugate pairs known as *pole pairs*. A zero of the function makes the numerator of $P(s)$ (and hence $P(s)$ itself) zero.

12. PREDICTING SYSTEM TIME RESPONSE FROM RESPONSE POLE-ZERO DIAGRAMS

A response pole-zero diagram based on $R(s)$ can be used to predict how the system responds to a specific input.

$$|R| = \frac{K \prod_z |L_z|}{\prod_p |L_p|} = \frac{K \prod_z \text{length}}{\prod_p \text{length}} \qquad \textbf{63.32}$$

$$\angle R = \sum_p \alpha - \sum_z \beta \qquad \textbf{63.33}$$

14. GAIN CHARACTERISTIC

- *quality factor, Q*

$$Q = \frac{\omega_n}{\text{BW}} \qquad 63.37$$

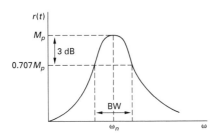

Figure 63.8 *Bandwidth*

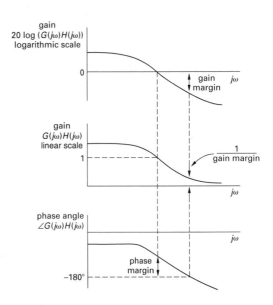

Figure 63.9 *Gain and Phase Margin Bode Plots*

16. STABILITY

The value of the denominator of $T(s)$ is the primary factor affecting stability. When the denominator approaches zero, the system increases without bound. In the typical feedback loop, the denominator is $1 \pm GH$, which can be zero only if $|GH| = 1$. Since $\log(1) = 0$, the requirement for stability is that $\log GH$ must not equal 0 dB.

A negative feedback system will also become unstable if it changes to a positive feedback system, which can occur when the feedback signal is changed in phase more than 180°. Therefore, another requirement for stability is that the phase angle change must not exceed 180°.

17. BODE PLOTS

Bode plots are gain and phase characteristics for the open-loop $G(s)H(s)$ transfer function that are used to determine the *relative stability* of a system.

The *gain margin* is the number of decibels that the open-loop transfer function, $G(s)H(s)$, is below 0 dB.

The *phase margin* is the number of degrees the phase angle is above −180° at the *gain crossover point*.

18. ROOT-LOCUS DIAGRAMS

A *root-locus diagram* is a pole-zero diagram showing how the poles of $G(s)H(s)$ move when one of the system parameters (e.g., the gain factor) in the transfer function is varied. The diagram gets its name from the need to find the roots of the denominator (i.e., the poles). The locus of points defined by the various poles is a line or curve that can be used to predict *points of instability* or other critical operating points. A point of instability is reached when the line crosses the imaginary axis into the right-hand side of the pole-zero diagram.

19. HURWITZ TEST

A stable system has poles only in the left half of the s-plane. The characteristic equation can be expanded into a polynomial of the form

$$a_0 s^n + a_1 s^{n-1} + \cdots + a_{n-1} s + a_n = 0 \qquad 63.39$$

The *Hurwitz stability criterion* requires that all coefficients be present and be of the same sign (which is equivalent to requiring all coefficients to be positive).

20. ROUTH CRITERION

The *Routh criterion*, like the Hurwitz test, uses the coefficients of the polynomial characteristic equation. The Routh-Hurwitz criterion states that the number of sign changes in the first column of the table equals the number of positive (unstable) roots. Therefore, the system will be stable if all entries in the first column have the same sign.

The table is organized in the following manner.

$$
\begin{array}{cccc}
a_0 & a_2 & a_4 & a_6 & \cdots \\
a_1 & a_3 & a_5 & a_7 & \cdots \\
b_1 & b_2 & b_3 & b_4 & \cdots \\
c_1 & c_2 & c_3 & c_4 & \cdots \\
\vdots & \vdots & \vdots & \vdots
\end{array}
$$

The remaining coefficients are calculated in the following pattern until all values are zero.

$$
b_1 = \frac{a_1 a_2 - a_0 a_3}{a_1} \qquad \textit{63.40}
$$

$$
b_2 = \frac{a_1 a_4 - a_0 a_5}{a_1} \qquad \textit{63.41}
$$

$$
b_3 = \frac{a_1 a_6 - a_0 a_7}{a_1} \qquad \textit{63.42}
$$

$$
c_1 = \frac{b_1 a_3 - a_1 b_2}{b_1} \qquad \textit{63.43}
$$

22. APPLICATION TO CONTROL SYSTEMS

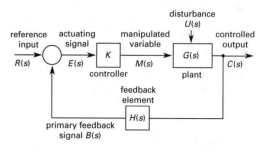

Figure 63.10 *Typical Feedback Control System*

Professional

EERM Chapter 66
Engineering Economic Analysis

> Chapter, section, equation, figure, and table numbers correspond to EERM. For additional study material, go to the corresponding chapter and section number in EERM.

11. SINGLE-PAYMENT EQUIVALENCE

- *future worth* (the equivalence of any present amount, P, at $t = 0$, to any future amount, F, at $t = n$)

$$F = P(1 + i)^n \qquad \text{66.2}$$

The factor $(1 + i)^n$ is known as the single payment *compound amount factor*.

- *present worth* (the equivalence of any future amount to any present amount)

$$P = F(1 + i)^{-n} = \frac{F}{(1 + i)^n} \qquad \text{66.3}$$

Index

3 dB point, 39

A

ABCD
  constants, for transmission lines, 54 (tbl)
  parameters, 52
Absolute index of refraction, 87
Absorption law, 16
Absorptive property, of Boolean algebra, 71
AC
  circuit fundamentals, 30
  generator, 45
  machinery, 56
  potential, in rotating machines, 56
  steady-state analysis, 37
  voltage, 30
Acoustic theory, 24
Action, mass, 61
Addition
  associative law of, 7
  commutative law of, 7
  of matrices, 7
  vector, 8
Admittance, 30 (tbl)
  transformer, 51
Advanced engineering mathematics, 17
Algebra, 6
  Boolean, 16, 71
  Laplace transforms, 14
  matrix, 7
Alternating current generators, 45
ALU (see "Arithmetic logic device")
AM, 90
  bandwidth, 90
  power, 90
American standard code for information
    exchange, 73
Ammeter, DC, 59
Amplifier, 61 (fig), 68
  differential, 68
  gain, 64
  noise, 69
  power, 64
Amplitude
  factor, 31
  modulation (see "AM")
Analog signal, 23
Analogy
  electric to magnetic, 28 (tbl)
  magnetic to electric, 28 (tbl)
Analysis
  asymmetrical fault, 50
  critically damped, 40
  fault, 49
  first-order, 39
  impedance, 37
  loop, 36
  node, 36
  of linear circuit, 33

overdamped, 40
second-order, 40
steady-state, 37
transient, 37, 38
underdamped, 40
unsymmetrical fault, 50
using Laplace transforms, 41
Analytic geometry, 10
AND logic, 74 (tbl)
Angle
  critical, 86
  direction vector, 8
  phase, 31
  small approximations, 10
  solid, 10 (fig)
Angstrom, unit conversion, 2 (tbl)
Antenna
  aperture, 89
  directivity, 88
  efficiency, 88
  electrical field, 87
  elemental, 87, 88
  height, 91
  ohmic resistance, 88
  power distribution, 88
  radiated power, 88
  radiation resistance, 88
  theory, 86
Aperture, of an antenna, 89
Apparent power, 32
  transformer, 51, 52
Approximations, small angle, 10
Architecture, computer, 73 (fig)
Arc length, 9 (fig)
Area
  by integration, 13
  unit conversion, 1 (tbl)
Arithmetic logic device, 79
Armature reaction, 45
ASCII, 73
Associative
  law of addition, 7
  law of multiplication, 7
  property, of Boolean algebra, 71
Asymmetrical fault analysis, 50
Augmented matrix, 7
Auxiliary field H, 27
Average
  power, 31
  value, 13, 31
  voltage, 59

B

Band-pass filter, 39 (fig)
Band-reject filter, 39
Bandwidth, 95 (fig)
Base-10 conversion, 15
Baseband signal, 90 (fig)
Base-b conversion, 15

Beam solid angle, 89
Bell-shaped curve, 15
BER (see "Bit error rate")
B-field, 26
BH curve (see "Magnetic hysteresis")
Bilateral Laplace transform, 18
Binary
  conversion, 16 (tbl)
  digits, 15
  number system, 15
Binomial coefficient, 15
Biot-Savart law, 21, 26
Bit, 15
  energy, 91
  error probability, 91
  error rate, 91
BJT (see "Transistor, BJT")
Bode diagram, 41
Bode plot, 41, 95
  gain margin, 95
  magnitude, 41
  methods, 43
  phase, 43
  phase margin, 95
  principles, 43 (fig)
  single pole at origin, 42 (fig)
  single pole on a negative real axis, 42 (fig)
  single zero at origin, 42 (fig)
  single zero on a negative real axis, 42 (fig)
Boolean algebra, 16, 71
  laws, 16
  theorems, 16
Bridge, Wheatstone, 59
Btu, unit conversion 1 (tbl)

C

Cable
  capacitance-graded, 49
  coaxial, magnetic field, 22 (tbl)
Calculation method, per-unit, 46
Calculus
  derivatives, 11
  integral, 13
Canonical representation, logic functions, 74
Capacitance, 20, 33
  defining equation, 33
  energy stored, 34
  -graded cable, 49
  of transmission line, 53
  s domain, 37, 41
  single-phase, 53
  steady-state, 37
Capacitive first-order circuit, 40
Capacitor, 20
  charge on, 19
  coaxial, 19
  parallel, 20, 34
  parallel plate, 20

Complete Preparation for the Electrical PE Exam

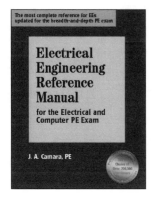

Electrical Engineering Reference Manual for the Electrical and Computer PE Exam

John A. Camara, PE

The **Electrical Engineering Reference Manual** is the most complete study guide available to engineers preparing for the electrical PE exam. It provides a clear, complete review of all exam topics, reinforcing key concepts with more than 300 solved example problems. The text is enhanced by hundreds of illustrations, tables, charts, and formulas, along with a detailed index. After you pass the PE exam, the **Reference Manual** will continue to serve you as a comprehensive desk reference throughout your professional career.

Practice Problems for the Electrical and Computer Engineering PE Exam: A Companion to the Electrical Engineering Reference Manual

John A. Camara, PE

This essential companion to the **Reference Manual** provides more than 440 practice problems with step-by-step solutions. You get immediate feedback on your progress and learn the most efficient way to solve problems.

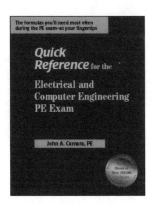

Quick Reference for the Electrical and Computer Engineering PE Exam

John A. Camara, PE

During the exam, it's vital to have fast access to the information you need to work problems. **Quick Reference** puts the important formulas at your fingertips, conveniently organized by subject.

Six-Minute Solutions for Electrical and Computer PE Exam Problems

John A. Camara, PE

Each of the more than 100 practice problems in **Six-Minute Solutions** is designed to be solved in six minutes or less—the average time per problem you'll have during the exam. Working the problems here will give you an edge on exam day.

Professional Publications, Inc.
1250 Fifth Avenue • Belmont, CA 94002
(800) 426-1178 • Fax (650) 592-4519
For secure, convenient ordering, try our complete online catalog at
www.ppi2pass.com